Environmental Pollution Control: Measures, Prevention and Safety Management

Edited by

Sanjeev Kumar
School of Hotel Management & Tourism
Lovely Professional University
Phagwara, Punjab, India

Mohammad Badruddoza Talukder
College of Tourism and Hospitality Management
International University of Business
Agriculture and Technology
Dhaka, Bangladesh

&

A.K. Haghi
Department of Chemistry
University of Coimbra
Coimbra, Portugal

Environmental Pollution Control: Measures, Prevention and Safety Management

Editors: Sanjeev Kumar, Mohammad Badruddoza Talukder and A.K. Haghi

ISBN (Online): 979-8-89881-489-2

ISBN (Print): 979-8-89881-490-8

ISBN (Paperback): 979-8-89881-491-5

Published by Bentham Science Publishers Pte. Ltd. Singapore, in collaboration with Eureka Conferences, USA.

First published in 2026.

BENTHAM SCIENCE PUBLISHERS LTD.

End User License Agreement (for non-institutional, personal use)

This is an agreement between you and Bentham Science Publishers Ltd. Please read this License Agreement carefully before using the ebook/echapter/ejournal (**"Work"**). Your use of the Work constitutes your agreement to the terms and conditions set forth in this License Agreement. If you do not agree to these terms and conditions then you should not use the Work.

Bentham Science Publishers agrees to grant you a non-exclusive, non-transferable limited license to use the Work subject to and in accordance with the following terms and conditions. This License Agreement is for non-library, personal use only. For a library / institutional / multi user license in respect of the Work, please contact: permission@benthamscience.org.

Usage Rules:

1. All rights reserved: The Work is the subject of copyright and Bentham Science Publishers either owns the Work (and the copyright in it) or is licensed to distribute the Work. You shall not copy, reproduce, modify, remove, delete, augment, add to, publish, transmit, sell, resell, create derivative works from, or in any way exploit the Work or make the Work available for others to do any of the same, in any form or by any means, in whole or in part, in each case without the prior written permission of Bentham Science Publishers, unless stated otherwise in this License Agreement.
2. You may download a copy of the Work on one occasion to one personal computer (including tablet, laptop, desktop, or other such devices). You may make one back-up copy of the Work to avoid losing it.
3. The unauthorised use or distribution of copyrighted or other proprietary content is illegal and could subject you to liability for substantial money damages. You will be liable for any damage resulting from your misuse of the Work or any violation of this License Agreement, including any infringement by you of copyrights or proprietary rights.

Disclaimer:

Bentham Science Publishers does not guarantee that the information in the Work is error-free, or warrant that it will meet your requirements or that access to the Work will be uninterrupted or error-free. The Work is provided "as is" without warranty of any kind, either express or implied or statutory, including, without limitation, implied warranties of merchantability and fitness for a particular purpose. The entire risk as to the results and performance of the Work is assumed by you. No responsibility is assumed by Bentham Science Publishers, its staff, editors and/or authors for any injury and/or damage to persons or property as a matter of products liability, negligence or otherwise, or from any use or operation of any methods, products instruction, advertisements or ideas contained in the Work.

Limitation of Liability:

In no event will Bentham Science Publishers, its staff, editors and/or authors, be liable for any damages, including, without limitation, special, incidental and/or consequential damages and/or damages for lost data and/or profits arising out of (whether directly or indirectly) the use or inability to use the Work. The entire liability of Bentham Science Publishers shall be limited to the amount actually paid by you for the Work.

General:

1. Any dispute or claim arising out of or in connection with this License Agreement or the Work (including non-contractual disputes or claims) will be governed by and construed in accordance with the laws of Singapore. Each party agrees that the courts of the state of Singapore shall have exclusive jurisdiction to settle any dispute or claim arising out of or in connection with this License Agreement or the Work (including non-contractual disputes or claims).
2. Your rights under this License Agreement will automatically terminate without notice and without the

Bentham Science Publishers Pte. Ltd.
No. 9 Raffles Place
Office No. 26-01
Singapore 048619
Singapore
Email: subscriptions@benthamscience.net

BENTHAM SCIENCE

CONTENTS

FOREWORD

Environmental pollution control has become a global concern and is inseparable from the policy objectives of governments around the globe.

There is a practical need to analyze this pressing issue concerning the relationship between environmental pollution and public health, and to adopt reasonable strategies to prevent health hazards and to follow up on research topics in this very important area.

However, many conventional waste pollution monitoring methods can lead to significant environmental issues, such as diminishing valuable resources, and cause a significant negative impact on society. This research-oriented edited collection introduces pollution treatment methods that can be adopted at local and international levels and examines appropriate resource management strategies for environmentally related issues. Presenting the latest research topics and innovative ideas to educate future citizens about the sustainable development of our planet.

The book investigates the diverse effects of environmental pollution and emphasizes how urgent it is to resolve the ecological costs of waste contamination. Each chapter blends academic theory, field research, case illustrations, and policy analysis, enabling a holistic understanding of the subject matter along with meaningful and productive experimental findings in this multidisciplinary field by renowned research scientists to report hazard analysis and human health risk assessment. This volume also reports practical achievements in environmental safety, environmental pollution control, along with technology innovation methods for better managing continuous screening and monitoring of chemical residues in environment.

Cristóbal Noé Aguilar
BBG-DIA, Bioprocesses & Bioproducts Research Group
Faculty of Chemical Sciences, Universidad Autonoma de Coahuila
Saltillo, Coahuila, Mexico

PREFACE

Environmental pollution control is very complex and multidimensional, and managing it effectively requires a range of advanced technologies. Environmental pollution is a global concern that is closely linked to the policy objectives of governments worldwide. It is crucial to analyze the relationship between pollution and public health and adopt practical strategies to prevent health hazards. Continued research on this important topic is also necessary.

This book covers important research topics in environmental pollution control and offers the most recent practical solutions by considering key concepts, a study plan, additional literature, and case studies. The authors of each chapter discuss the limitations of the current measures, prevention, and safety management, and suggest complementary treatment methods.

This edited collection is creating an interdisciplinary platform for researchers and practitioners to present and discuss the most recent innovations, trends, and concerns, as well as the practical challenges encountered and solutions adopted in the field of environmental science. It brings to the fore the case studies and practical relevance of a framework for safety management and the future of sustainability in the ecosystem.

Sanjeev Kumar
School of Hotel Management & Tourism
Lovely Professional University
Phagwara, Punjab, India

Mohammad Badruddoza Talukder
College of Tourism and Hospitality Management
International University of Business
Agriculture and Technology
Dhaka, Bangladesh

&

A.K. Haghi
Department of Chemistry
University of Coimbra
Coimbra, Portugal

List of Contributors

Amanjeet Bhalla	Chitkara College of Hospitality Management, Chitkara University, Rajpura, Punjab, India
Aman Bharadwaj	University Institutes of Tourism and Hospitality Management, Chandigarh University, Mohali, Punjab, India
Amrik Singh	School of Hotel Management and Tourism, Lovely Professional University, Jalandhar, Punjab, India
Arshad Bhat	Amity Institute of Liberal Arts, Amity University, Mumbai, Maharashtra, India
Durgeshwary Kolhe	Department of Clinical Psychology, School of Vedic Sciences, MIT-ADT University, Pune, Maharashtra, India
Farhana Yeasmin Lina	Department of Tourism and Hospitality Management, Daffodil International University, Dhaka, Bangladesh
Firoj Kabir	Department of Tourism and Hospitality Management, Daffodil International University, Dhaka, Bangladesh
Gaurav G. Pawar	Department of Economics, Abasaheb Marathe Arts and New Commerce, Science College Rajapur, Ratnagiri, Maharashtra, India
Geetha Manoharan	School of Business, SR University, Warangal, Telangana, India
Ghanshyam Shakar	School of Engineering & IT, MATS University, Raipur, Chhattisgarh, India
Harshad L. Jadhav	Department of Economics and Research Centre, Karmaveer Bhaurao Patil College, New Mumbai, Maharashtra, India
Kamarun Muhsina	Department of Tourism and Hospitality Management, Daffodil Institute of IT, Dhaka, Bangladesh
Manish Baboo Agarwal	Sharda School of Basic Sciences and Research, Sharda University, Agra, Uttar Pradesh, India
Manish Kumar Panday	Department of Applied Sciences, Anand Engineering College, Keetham, Agra, Uttar Pradesh, India
Manu Mehrotra	Sharda School of Basic Sciences and Research, Sharda University, Agra, Uttar Pradesh, India
Md. Afzal Hossain	College of Business Administration, International University of Business Agriculture and Technology, Dhaka, Bangladesh
Md. Atikur Rahman	School of Management, Jiujiang University, Jiujiang, China
Md. Mamonur Rashid	Department of Business Administration, Daffodil Institute of IT (DIIT), Dhaka, Bangladesh
Md. Omar Faruk	Department of Business Administration, Daffodil Institute of IT (DIIT), Dhaka, Bangladesh
Mohammad Badruddoza Talukder	College of Tourism and Hospitality Management, International University of Business Agriculture and Technology, Dhaka, Bangladesh
Pankaj Mittal	Department of Applied Sciences, Anand Engineering College, Keetham, Agra, Uttar Pradesh, India

Parambir Singh	Chitkara College of Hospitality Management, Chitkara University, Punjab, India
Priyanka P. Pandey	School of Hotel Management & Tourism, Lovely Professional University, Phagwara, Punjab, India
Sanjeev Kumar	School of Hotel Management & Tourism, Lovely Professional University, Phagwara, Punjab, India
Santosh Kumar Bhardwaj	Department of Mathematics and Computing, Madhav Institute of Technology & Science, Gwalior, Madhya Pradesh, India
Seema Agarwal	Department of Mathematics, SRM Institute of Science and Technology, Ghaziabad, Uttar Pradesh, India
Shweta Dewangan	Faculty of Commerce, ICFAI University, Raipur, Chhattisgarh, India
Siddhartha Dey	University Institutes of Tourism and Hospitality Management, Chandigarh University, Mohali, Punjab, India
Sohel Rana	College of Business Administration, International University of Business Agriculture and Technology, Dhaka, Bangladesh
Srishti Shaifali Minz	Govt. Rajmohini Devi Girls P.G. College, Sarguja University, Ambikapur, Chhattisgarh, India
Wasswa Shafik	School of Digital Science, Universiti Brunei Darussalam, Bandar Seri Begawan, Brunei Darussalam Dig Connectivity Research Laboratory (DCRLab), Kampala, Uganda

CHAPTER 1

Impact of Environmental Pollution on Food Security

Kamarun Muhsina[1,*], **Mohammad Badruddoza Talukder**[2] and **Farhana Yeasmin Lina**[3]

[1] *Department of Tourism and Hospitality Management, Daffodil Institute of IT, Dhaka, Bangladesh*

[2] *College of Tourism and Hospitality Management, International University of Business Agriculture and Technology, Dhaka, Bangladesh*

[3] *Department of Tourism and Hospitality Management, Daffodil International University, Dhaka, Bangladesh*

Abstract: Environmental pollution is becoming a significant threat to global food security, with a profound impact on food availability, quality, and stability. This chapter aims to present how various types of pollution, including air, dust, water, biological, and climate change, impact food production systems. These pollutants affect the quality of food harvested from poisonous soil and water. Pollution also contaminates water, disrupts water-based food chains, and destroys crops due to air pollution. This theory acutely reveals the interconnection between nature, pollution, and food systems, highlighting how pollution at one point affects the entire food supply chain. The chapter also investigates the exact impacts of crop harvesting and shows how pollution contributes to reduced nutritional value, bringing theious economic consequences to the farming communities.

Furthermore, it highlights the differential impacts of pollution on food security across various socioeconomic backgrounds, underscoring the need for location-based strategies to address these challenges. Fisheries and cattle are mostly vulnerable to pollution that accumulates through the food chain. Moreover, both food availability and human health are also at risk. The chapter discusses moderation strategies and deliberates on sustainable agricultural practices that set pollution control policies and technological solutions to protect the human food system. Lack of harmonized international hard work and environmental fairness considerations, as pollution and its impacts on food security often excessively affect vulnerable populations. In conclusion, this chapter aims to integrate science-driven solutions to address future challenges, such as pollution and implementation, thereby enhancing food and food security in a changing world.

Keywords: Challenges, Climate, Environment, Food, Impact, Pollution, Security.

* **Corresponding author Kamarun Muhsina:** Department of Tourism and Hospitality Management, Daffodil Institute of IT, Dhaka, Bangladesh; E-mail: kamarunmuhsina7@gamil.com

Sanjeev Kumar, Mohammad Badruddoza Talukder & A.K. Haghi (Eds.)

INTRODUCTION

Pollution of the environment is a threat to global food security, as it affects the capacity of individuals, communities/and nations to access protective, nutritious, and adequate food at any given time. Pollution is also closely linked to and impacts the food system, particularly in relation to the development of sustainable agriculture and food systems, food security, and the health of ecosystems. (Bisht, Kamboj *et al.*, 2020). For example, water, land, and even the air become contaminated with heavy metals, pesticides, and microplastics, which reduce crop production rates and affect livestock health. In addition, pollution disrupts ecosystems, contributing to their degradation and ultimately compromising food systems.

As the global population increases, competition for resources intensifies, exacerbating the already growing threats from pollution. These are compounded by climate change, which is caused by GHG emissions, affecting weather patterns, increasing the incidence of extreme events, and hastening soil degradation (Ramanathan & Feng, 2009). For poor nations in the developing world, where dependence on food production from the agricultural sector is widespread, the impact is significantly more damaging, exacerbating the dangers posed by hunger, malnutrition, and the cycle of poverty.

Defining Environmental Pollution

The act of putting a deleterious substance or energy into an environment that is unsuitable for the specific purpose in a manner that results in a detrimental alteration of the nature of the entirety of the climate; a decrease in quality or deterioration of the features of an environment; negative effects on living species. Such contaminants can be grouped according to their source, phase state, or effects: chemical (industrial effluents, pesticides, metals), physical (plastics, noise, radiation), and biological (pathogens, alien species). Pollution can be either anthropogenic in origin, as observed in industrialization or deforestation, or natural, resulting from processes such as volcanic eruptions, basin desertification, or intensive and uncontrolled agricultural activities (Bisht, Kamboj *et al.*, 2020). Some of these pollute and alter major environmental features, including air, water, soil, and life forms, with knock-on effects on global systems such as weather and food production. Besides affecting the environment, pollution affects people's health both by causing and exacerbating respiratory disorders, waterborne diseases, and other illnesses resulting from exposure to toxins. It also exacerbates socioeconomic injustice because polluting activities affect individuals living in the vicinity of polluted sites or those who receive inadequate amounts of clean resources (Brehm & Pellow, 2022). Every problem of environmental pollution is best solved through the use of science, policy, technology, and community involvement. To address these concerns, this paper presents a comprehensive view of the problem, an analysis of its impacts, and approaches to ecosystem rehabilitation and the advancement of sustainable development goals.

Overview of Food Security

Food security is defined as the affordable access by all people at any one time to the required amounts of the appropriate kinds of food (Pinstrup-Andersen, 2009). It is a multidimensional concept encompassing four key dimensions: availability, accessibility, use, and stability. Stocking establishes the qualitative and quantitative availability of food through production, trade, and distribution systems. Accessibility is defined as the economic and physical ability to gain access to food and individual and community resources. In this case, biological utilization of food products is promoted by focusing on nutrition, safe food intake, and healthcare to ensure the optimal conversion of food into energy and health. Stability refers to a country or geographical region having a steady food supply, access to food, and effective food utilization in the face of shocks such as conflicts, climate variations, and economic hardships (Pingali, Alinovi *et al.*, 2005). These dimensions point towards sustainable and equitable systems to achieve food security on a global level.

Linkages Between Pollution and Food Systems

Pollution affects agricultural and food production systems in the broadest sense, as well as the environment from which these production systems derive their sustainability. These are basic natural resources that support food production, and they include soil, water, and air, which are unproductive or damaged, making food production and resilient food systems almost impossible. The processes of chemicalization, mechanicalization, and industrialization have contaminated the soil with industrial waste, excessive pesticides, and heavy metals, resulting in diminished soil fertility, reduced agricultural yield, and compromised food quality (Rathod, Saras *et al.*, 2025). Contamination from agricultural effluent, chemical disposal, and raw sewage remains the primary threat to the quality of water used in supporting agricultural productivity, irrigation, and human consumption. Likewise, air pollution affects several factors, including the levels of greenhouse gases and particulates, as well as moderating factors such as temperature, moisture, and radiation, which impact crop production (Ramanathan & Feng, 2009). In particular, total BFRs can accumulate along the food chain, resulting in damaging consequences for consumers and compromising food safety. These disruptions demonstrate how closely linked the demands for food provision are with the need for sustainable pollution and environmental conservation (Prabhakar, Singh *et al.*, 2012).

Objectives

The objectives of this chapter are given below:

1. To identify environmental pollution and its interconnections with food security.
2. To explore various types of environmental pollution and its impact on food security.
3. To examine the influence of pollution on livestock farming.
4. To highlight challenges and set some strategies for reducing the impact of pollution on food security.
5. To recommend some actionable steps for policymakers to address gaps in environmental pollution and food security.

LITERATURE REVIEW

Types of Environmental Pollution Affecting Food Security

The harmful effects of environmental pollution also threaten food security, as it adversely affects the production and distribution of ecosystems (Richardson, 2010). Air, soil, water, and chemical pollution affect all the necessities required for growing, refining, and marketing food. These pollutants impact soil quality, contaminate water sources, and alter atmospheric conditions, which in turn affect crop production, quality, and accessibility.

Air Pollution

Sulfur dioxide, nitrogen oxides, and particulate matter reduce crop productivity and contaminate food throughout various stages of growth and preparation.

Soil Pollution

Soil pollution by heavy metals, pesticides, and industrial waste reduces soil fertility and pollutes the food chain (Ariyo *et al.,* 2020).

Water Pollution

Contamination of water affects the water supply for irrigation and can deposit pathogens, heavy metals, and microplastics into crops and fish.

Chemical Pollution

Excessive use of fertilizers and pesticides has some adverse effects on food that contains chemicals, which are hazardous to health.

Climate Change and Greenhouse Gas Emissions

Global warming, which is a consequence of climate change, influences the onset of droughts, floods, and other extreme events that have a direct impact on food production (Rosenzweig, Iglesius *et al.,* 2001).

Pathways of Pollution Impact on Food Security

Pollution, in its most comprehensive definition and across its various dimensions, jeopardizes food security both directly and indirectly by influencing the elements of availability, access, and quality of human food (Acevedo, 2011). These routes include contaminating and degrading vital natural resources, such as soil, water, and air, which are essential for supporting agricultural and food production systems. It can penetrate the bodies of crops through the soil, travel through irrigation water sources from one crop to another, and fly into the air, which lowers crop yields and food quality. Second, Chemical pollutants and toxins, which largely include pesticide residues and heavy metals, accumulate in the food chain, posing a risk to human well-being and depleting many life-supporting systems (Sonone, Jadhav *et al.*, 2020). These pathways have revealed that pollution is multifaceted; its capability to alter access, reliability, and consumption of food resources cannot be overstated. Understanding these mechanisms is considered necessary to enhance the overall and effective measures for the prevention of pollution and for food security.

Pollutant Transfer through Soil to Crops

Soil is a basic form of agricultural produce; the pollution of which by toxicants such as heavy metals, industrial effluents, pesticides, and other dangerous compounds threatens food safety (Oladoye, Olowe *et al.*, 2022). Such pollutants are absorbed by crops through their root systems, causing a buildup of toxic substances within the plant tissues. These processes not only decrease crop yields but also reduce the nutritional value of the food obtained from those yields. Soil-borne pollutants are known to reduce crop yields, stunt crop growth, and even produce toxic crops when ingested. Furthermore, current and past pollution reduce the future productivity of agricultural land, which complicates existing problems for farmers and global food security (Hossain, Krupnik *et al.*, 2020).

Water Contamination and Irrigation Issues

Water pollution is closely related to food security, mainly because polluted water is used for irrigation in food production. Industries release deleterious materials into water bodies, while fertilizers and pesticides from agricultural sources pollute the water, and the unrestricted discharge of untreated urban waste contaminates the soil and crops. These substances include nitrates, phosphates, and pathogens, which cause diseases in crops, lower productivity, and generate hazardous contaminants in foodstuffs. In addition, the use of low-quality water for irrigation lowers soil quality, structure, and productivity, posing long-term issues in agriculture (Singh, 2021). Since water sources that supply irrigation systems are contaminated, this poses a significant danger to food handlers, those who consume the produce, and the farmers themselves.

Airborne Pollutants and Crop Yield Reduction

Ground-level ozone, particulate matter, Sulphur dioxide, and nitrogen oxides are among the most deleterious pollutants affecting crop yields and quality. Similarly, ozone enters mesophyll cells through open stomata, causing oxidative damage that leads to reduced photosynthesis and plant growth (Grulke & Heath, 2020). The particulate matter that accumulates on plant surfaces prevents light from penetrating, thereby harming plants' photosynthesis. Meanwhile, herbivores and saprotrophic animals are affected by sulfur and nitrogen oxides, which cause acid deposition, changes in individual soil pH levels, and reduced nutrient availability. These pollutants, when combined, decrease crop yields and quality, jeopardizing food security and the food economy for agricultural societies (Mihai, Gündoğdu *et al.*, 2021).

Pesticides, Heavy Metals, and Food Chain Contamination

The misuse of pesticides and the adoption of heavy metals in agriculture have a great potential for bioaccumulation in the food chain. Chemicals applied in pest control, at their best, cause residue deposits on crops, and at worst, are washed down into soil and watercourses, affecting non-target species and human beings (Ali, Ullah *et al.,* 2021). To the same effect, lead, cadmium, mercury, and other related heavy metals are absorbed by crops and can even be magnified as animals feed on crops produced from these contaminated soils. This bioaccumulation interferes with numerous ecosystems, resulting in declines in species numbers, while the risk to humans who consume these contaminated food products increases, leading to chronic diseases. Such disruptions introduce vulnerabilities into the food regulatory system, requiring heightened monitoring and risk-reduction measures.

Impacts on Crop Yield and Quality

The abuses of pesticides and the adoption of heavy metals in agriculture have a great potential for bioaccumulation in the food chain. Chemicals applied in pest control, at their best, cause residue deposits on crops, and at worst, are washed down into soil and watercourses, affecting non-target species and human beings (Ali, Ullah *et al.,* 2021). To the same effect, lead, cadmium, mercury, and other related heavy metals are absorbed by crops and can even be magnified as animals feed on crops produced from these contaminated soils. This bioaccumulation interferes with numerous ecosystems, resulting in a decline in species numbers, while the risk to humans who consume these contaminated food products increases, leading to chronic diseases. Such disruptions introduce vulnerabilities into the food regulatory system, requiring special attention to monitoring and risk reduction measures.

Reduction in Agricultural Productivity

Tackling agricultural productivity and food security depends on reducing environmental pollution, particularly at the levels of water and air (Zahoor & Mushtaq, 2023). Crops are negatively impacted by airborne pollutants, such as ground-level ozone and particulate matter, which interfere with and reduce their ability to produce energy. For example, ozone can damage leaves through oxidative stress, reducing chlorophyll levels and thereby decreasing the plant's capacity for photosynthesis. Water pollution resulting from industrial discharges, pesticides, and fertilizers also pollutes sources where water used in field irrigation is contaminated due to a reduction in water quality and an increase in soil salinity (Okorogbona, Denner *et al.*, 2018). Eventually, this contamination can reduce the soil's fertility, resulting in lower crop yields. These impacts not only harm immediate agricultural production but also have long-term implications for soil health, making such crops susceptible to reduced productivity over time, even in the future. Significant aspects of the economy include higher costs for farmers' products, resulting in the need for more inputs (such as fertilizers and pest control), and potential financial losses due to reduced harvests.

Nutritional Deficiencies in Polluted Crops

Polluted crops can become nutritionally deficient, which can have significant negative impacts on both crop quality and human health benefits (Aulakh & Malhi, 2005). Potentially heavy metals include lead, cadmium, and arsenic, the main ones found in polluted soil and water, which can be taken up by plants and stored in edible plant parts. Essential nutrients such as calcium, magnesium, and iron, for example, can be hindered by this contamination, resulting in crops that lack these important elements. Pollutants can also produce deleterious effects on the plant's ability to synthesize proteins and other vital compounds necessary for normal growth (Nagajyoti, Lee *et al.*, 2010). Moreover, pollutants can damage a crop's microbiome—the natural process in soil that helps plants efficiently absorb nutrients. Not only does this compromise the nutritional needs of consumers who rely on these crops for their nutrition, but it also reduces the overall nutritional value of the food supply. As a result, pest species may lead to higher rates of malnutrition and other related health issues in people, especially where farming products are the main source of food.

Economic Consequences for Food Production

The effects of setting viable yields and the diminished nutritive value in contaminated crops cause major economic imbalances between the growers and users (Vurro, Bonciani *et al.*, 2010). For farmers, low productivity results in a decrease in the income they generate from their farms. This leads to a cycle where, due to low yields, they are unable to earn enough to purchase the tools, seeds, or fertilizers required to ensure or enhance future yields. Furthermore, the costs of penalties for polluting the environment, such as the need to install water purifiers or adopt costly farming methods, directly reduce farmers' revenues. From the consumer side, poor yields and contaminated produce result in higher food prices because an inadequate supply cannot meet demand. This price inflation has an adverse effect, as it disproportionately impacts low-income communities that typically spend a high proportion of their budgets on food, leading to food insecurity and a higher level of poverty (Carney, 2012). All in all, the costs associated with agricultural decline due to pollution hurt a country's economic performance and may lead to an increased influx of food imports.

POLLUTION IMPACTS ON FISHERIES AND AQUATIC FOOD SOURCES

Water pollution is one of the significant threats to fish stocks and aquatic food sources, causing severe impacts on habitats and reducing the growth and health of water bodies (Bukola, Zaid *et al.*, 2015). Aqua systems are critical for sustaining species diversity, in addition to being a central source of food and income to millions of individuals globally. In addition to affecting the health and development of fish and other aquatic organisms, these contaminants are also known to degrade the aquatic environment and the species that inhabit it. That is not just an ecological problem but also an economic and public health issue for those societies dependent on fish production, especially through fishing activities (McClanahan, Allison *et al.*, 2015). Growing pollution from industrial, agricultural, and urban sources makes it crucial to assess its effects on aquatic food resources and to develop sustainable measures for protecting these vital resources for future generations' fish supply. Here is a detailed overview of how pollution impacts fisheries and aquatic food sources:

Harm to Aquatic Species and Ecosystems

Ecological pollution directly affects waterborne life and, in many cases, leads to the destruction of the affected ecosystems. Heavy metals, such as mercury, lead, and cadmium, as well as industrial chemicals found in industrial effluents, can enter the water and bioaccumulate and biomagnify when consumed by fish and other seafood. Acute stress can lead to adverse health effects, including internal organ damage, impaired reproduction, and mortality in fish (Schreck, 2010). Many contaminants have an impact on the endocrine systems of water-dwelling organisms, causing developmental distortions and reduced population densities, such as pesticides and pharmaceuticals. Some threats, such as eutrophication resulting from high nutrient concentrations from agricultural activities, contribute to the excessive growth of algae in the water body. These algal overgrowths can hinder penetration of light into the water, reduce the productivity of the oxygen content in the water, and form what can be referred to as 'dead areas' where the majority of marine life cannot exist, harming entire ecosystems (Bailey, Meyer *et al.*, 2020)

Decline in Fish Populations and Fisheries Yields

Pollution is a factor that leads to the reduction of fish stocks, thereby impacting fish production, which in turn affects the entire population that depends on them. Pollutants dissolved in water decrease the number of fish through increased mortality and hinder reproduction (Hamilton, Cowx *et al.*, 2016). This consequently results in a reduction of the available fish for harvest, hence a reduced yield in fisheries. Additionally, most fish are intolerant to changes in the quality of the water in which they live; even low concentrations of pollutants can be detrimental to fish, as they affect breeding and feeding grounds. Fish population decline can lead to economic problems for fishermen and fishing-dependent societies, as well as increased food shortages, particularly in protein, among populations that rely on fish as their primary source of food.

Impact on Aquatic Food Chain and Biodiversity

Pollution affects the water chain, from producers to top consumers of the food chain, such as phytoplankton and zooplankton, respectively. For example, nutrient pollution creates algal blooms—the growth of algae—that may produce toxins fatal to organisms such as fish larvae and shellfish. When such organisms cannot survive, it impacts the food chain to the larger fish and even the predators that feed on them. Pollution causes the deterioration of habitats, such as coral reefs and mangroves, through sedimentation or exposure to chemicals, which reduces the ecosystem's biodiversity and its ability to recover. The loss of this species further threatens the sustainability of both the populations of the different species involved and fisheries that depend on a diverse species for productivity and sustainability (Lapointe, Cooke *et al.*, 2014).

Public Health Risks

Raw fish and other aquatic products that are contaminated with pollutants pose a threat to society's health and innovation. Various chemical pollutants, including heavy metals, which are toxic to humans, may be stored in fish tissues. For example, mercury, when ingested, becomes dangerous to the development of the neurological system in pregnant women and children below 15 years old. The intake of POPs and plastics can lead to health complications, including cancer, immune system issues, and endocrine disorders, among other ailments, in marine organisms (Okoye, Addey *et al.*, 2022). Furthermore, nutrient pollution that supports the growth of algae results in the formation of these blooms, which are contaminated with toxins that are hazardous when ingested directly or after filtering through shellfish, causing poisoning and severe health conditions among people.

Economic Implications for Coastal and Fishing Communities

The economic effects of pollution on fisheries are unfathomable. Fishing associated with small boat operators is one of the sectors most severely affected by the impacts resulting from dwindling fish numbers and poor water quality. Pollution eradication measures, including the purchase of advanced technologies, cleaning up polluted environments, and restoring the health of affected ecosystems, come with a cost that is borne by local and national fiscal systems (Ajibade, Adelodun *et al.*, 2021). Additionally, reduced fish stock affects the fisheries production budget; consumers are left with fewer affordable retail fish options as costs rise due to declining fish production. In our case, some of the regions that will be negatively affected include those whose populace relies on the fishing industry for their employment needs, coastal businesses whose earnings will be affected through low tourist arrivals since the waters will be empty, or even regions whose general economy will be unstable due to the impacts of conservation measures. In the long run, this economic weakness gives rise to communal and social issues that are pressured to transform or be addressed due to current fishing conditions.

Challenges in Pollution Management and Mitigation

Measures taken to prevent and control pollution in the bodies of water pose certain difficulties. By this, the author means that industrial, agricultural, and urban runoff are, most of the time, massive and contain a vast cocktail of pollutants, which makes source control solutions difficult or impossible (Council, Earth *et al.*, 2009). This is mainly because policies have been implemented to control pollution through regulations, and new technologies introduced may lack the necessary enforcement and overall compliance. Secondly, many water bodies are international, thus making it very difficult for countries to cooperate in controlling the pollution of water sources. Much as contemporary policies focus on policy changes, the enhanced use of technology, and community engagement, there is a need to adopt a multidimensional approach in addressing the issue of pollution that affects fisheries and sources of fish and other aquatic foods. To prevent further damage to these essential resources, Mi [2] Me [2] strategies, such as enhancing the functionality of wastewater treatment plants, introducing methods that produce food while minimizing water use, and improving the effectiveness of industrial regulatory bodies, should be adopted (Scott, Faruqui *et al.*, 2004).

These specifics highlight the importance of taking crucial steps to mitigate the environmental and food security implications of pollution on fishing and the supply of fish resources, as well as the positive impacts on global economies.

POLLUTION AND LIVESTOCK FARMING

Pollution poses numerous challenges for livestock farming, making it more difficult to maintain animal health, achieve optimal productivity, and produce high-quality food. At the same time, livestock farming not only contributes to pollution but also suffers from it, through exposure to polluted air, contaminated water, and soil (Weldeslassie, Naz *et al.*, 2018). Solid and liquid waste, including manure, releases harmful airborne pollutants like ammonia and methane. These gases not only degrade air quality but also contribute to climate change as powerful greenhouse gases. In runoff, chemicals such as antibiotics, hormones, and nutrients from feedlots can pollute a region's water supply, while also affecting local ecosystems. Contamination of soil can occur through improper disposal of waste or the application of pesticides, which is detrimental to the overall productivity of the land and to livestock grazing on it (Kraham, 2017). This pollution poses risks not only to animals and farms but also to human health through contaminated food. Addressing these challenges effectively is crucial for the future of livestock farming, public health, and environmental sustainability.

Here is a detailed breakdown of each point on how pollution impacts livestock farming.

Soil and Feed Contamination

It emphasized that contaminated soil and water pose a danger to livestock, which in turn affects animal health (Lu, Song *et al.*, 2015). When water sources used for irrigation are contaminated with chemicals such as heavy metals, pesticides, or industrial waste, and livestock feed crops are irrigated with this water, the toxins accumulate in the bodies of these animals. Thus, soil contaminated with heavy metals, chemicals, or humus can lead to feed crops with high levels of pollutants. When livestock consume contaminated feed, the toxins pass through the gastrointestinal tract and reach the animal's tissues and organs, resulting in stunted growth, reproductive failure, and poor health. Second, soil pollution alters nutrient availability in pastures by reducing nutrient uptake, thereby creating or likely creating a shortfall in nutrient availability for livestock (Masters, Norman *et al.*, 2019). Many of these contaminants pose significant risks to the health of the animals, as well as the quality of meat and dairy products derived from them.

Health Impacts on Livestock

Impurities arising from contaminated feed can directly cause diseases in livestock, leading to productivity declines due to sickness. Many of these and other pollutants, such as heavy metals, pesticides, and industrial chemicals, can, in fact, lower the immune systems of the livestock (Gupta, Bandyopadhyay *et al.*, 2021). For example, heavy metals are poisonous to organs and can lead to neurological sicknesses, pesticide scums, disrupt hormones, and have an effect on the reproductive system. Furthermore, impurities present in the feed cause alimentary disorders, resulting in output depletion and decreased growth rates. From a farmer's perspective, this not only leads to reduced meat and dairy output but also to related health issues, which in turn elevate veterinary bills. Polluted animals are often susceptible to diseases and illnesses, resulting in a higher mortality rate, which in turn affects the farm's viability (Hu, Cheng *et al.*, 2017).

Transmission of Pollutants through Animal-based Food

Any human being can be affected by pollutants through the consumption of contaminated animal-based foods (Lei, Zhang *et al.*, 2015). Different contaminants can deposit in the muscles, liver, milk, and eggs of animals when livestock feed is contaminated. Substances such as heavy metals, pesticides, and POPs are transferred to consumers through meat, dairy, or eggs. For instance, mercury and cadmium, two naturally occurring elements that are often present in water and feed, can accumulate in animal tissues and cause various diseases in humans. Research done on the effects of such toxins over time has shown that they cause various ailments, neurodegenerative disorders, developmental complications in children, and some types of cancer. Bioaccumulation of pollutants through the food chain may also contribute to the spread of antimicrobial resistance when antibiotics are contained in animal feed that is consumed (Sources, Koutsoumanis *et al.*, 2021). This can defeat the purpose of administering antibiotics to human infections, thus posing major public health problems. The health of both animals and people is at risk as long as the feed and water for animals are toxic.

REGIONAL AND GLOBAL CASE STUDIES

The effects of pollution on food security are illustrated in this paper through various regional and global examples, spanning different continents. Rice paddies in Asia are severely affected by contamination and heavy metal pollution, including arsenic and cadmium, in the water and soil, resulting from industrial effluent discharge and pesticide application. This pollution not only affects food productivity in the form of a reduction in rice yield, but it is also hazardous to the health of populations that depend on rice as their primary food source (Asma, Subrahmanyam *et al.*, 2023). In Africa, Water pollution from industrial waste discharge, agricultural drains, and improper sewage disposal has a sharp impact on drinking water, irrigation/fisheries, and, consequently, food supplies. Adeyemo (2003) explained that rivers and lakes that provided support to agriculture and fishing industries are all polluted, which has reduced food production and forced the fishing industries to adapt or shut down. North America also faces its own problems with air pollution; industrial and automobile emissions release gases that reduce crop yields in agricultural areas. Primary pollutants, such as ground-level ozone, can interfere with photosynthesis and reduce the yields of key food crops, resulting in significant losses for farmers. These situations highlight how socioeconomic factors, including levels of development, play a crucial role in shaping the impact and response to environmental challenges. These factors influence the extent and consequences of pollution on food security through development policies and standards, as well as the affordability of technologies. Some developed countries can afford the measures and changes required to tackle pollution, while many developing countries cannot; as a result, the latter are more likely to be affected. Such cases necessitate prompt and diverse regional solutions to combat pollution and safeguard global food security (Sundram, 2023).

Comparative Analysis of Pollution Impact on Food Security

A cross-country comparison of the effects of pollution on food security reveals that social and economic contexts are the most influential factors in determining the level and type of these effects (Richardson, 2010). The impact of pollution on agriculture and food supplies is, however, often offset in wealthier industrialized nations by the ability to manage pollution through the implementation of various technologies, along with stringent environmental laws. Here are the details:

a. **Economic Capability and Pollution Regulation**:
    - It shows that relatively richer countries may have the means to implement environmental laws and to fund technologies that help agriculture overcome the losses from pollution.
    - Sophisticated infrastructure and policies in place can reduce pollution levels, making food production more secure (Lu, Jenkins *et al.*, 2015).
    - Advanced infrastructure and policies can help mitigate pollution and protect food production, ensuring greater food security (Lu, Jenkins *et al.*, 2015).
b. **Continued Challenges in Developed Nations**:
    - Nevertheless, there is a population that is aware of their industrial and vehicular emissions, and they are capable of reducing crop yield and increasing the cost of food production.
    - Sustainable economic investment fosters food system resilience, but it also creates environmental threats, even in the most developed countries. The tension between economic growth and environmental health can lead to vulnerabilities in the food system, even in economically advanced countries.
c. **Impact on Low-income Countries**:
    - Developing countries with limited capacity to generate resources for investment and inadequate formulation and implementation of environmental laws are grappling with severe food insecurity challenges, primarily due to pollution (Ajibade, Adelodun *et al.*, 2021).
    - Industrial pollution, pesticide use, and improper waste handling harm the water, soil, and air, which in turn damages agriculture. (Elodun *et al.*, 2021).
    - Unregulated industrial activities, pesticide use, and inadequate waste management contribute to water, soil, and air pollution, severely affecting agriculture.
d. **Adaptability and Infrastructure**:
    - Developed countries have the facilities needed to detect and contain pollution, and are thus in a position to adapt rapidly to the different conditions prevailing in the environment.
    - Whereas lower-income regions generally lack the capacity or flexibility to afford the loss of food production over a longer term (Kakaei, Nourmoradi *et al.*, 2022).
e. **Food Import Dependence**:
    - Developed countries can import food from other countries, allowing them to feed their populations, even if pollution affects their own farming.

- Compared to the global average, inhabitants of less-developed countries, who depend mostly on local farming, are more vulnerable to worsening of these indicators in the event of increased pollution (Khatri, Kumar *et al.*, 2024).

f. **Socioeconomic Disparities**:

- The community's reactions towards pollution impact depend on socioeconomic conditions. Some regions can afford to implement new technologies and improve practices, while others may still not realize the need to contain pollution at all.
- They make it clear that specific international assistance and collaboration at this critical juncture will be required to mitigate and intervene on behalf of endangered communities, as well as to encourage and support more sustainable approaches to agriculture.

These disparities underscore the need for targeted international support and cooperation to safeguard vulnerable populations and foster sustainable agricultural practices.

g. **Need for Tailored Approaches**:

- Regional solutions must consider the socioeconomic disparities between different regions in terms of investment in cleaner technologies, the enforcement of stricter environmental standards, and the sustainable development of communities most affected by pollution (Camagni, Capello *et al.*, 1998).

MITIGATING POLLUTION TO IMPROVE FOOD SECURITY

Preventing pollution must remain a critical agenda for the desired development of enhancing food security, as pollution weakly affects agriculture and consumable foods. Pollutants negatively affect the food system by contaminating manufactured crops, reducing crop production rates, and impacting food production rates for livestock, as well as compromising food safety (Sharma, Adhoni *et al.*, 2023). Effective pollution control also aids in the conservation and improvement of the environment, especially in supporting and sustaining food production for a growing population. Some other measures that can be taken to reverse pollution are: official measures, for instance, clean industrial processes and coming up with stringent regulations and policies concerning the environment, reverse approaches such as encouraging better farming practices among farmers, and measured structures, for instance, here overhaul of waste management systems (Smith & Siciliano, 2015). As pollution control is receiving increasing importance from governments, industries, and local communities, it is possible to look forward to a safer food future.

Policies for Reducing Environmental Pollution

Policies play a crucial role in preventing pollution, managing the environment, and ensuring food security. Procedures such as establishing laws governing the emission of gases from industries, the disposal of waste, and the minimal use of toxic chemicals are sensible measures. For instance, legislation that addresses the emission of greenhouse gases or prohibits the use of certain pesticides can directly enhance specific crop ecosystems, including those related to air and soil quality. The second is environmental regulatory policies that press industries to adopt cleaner technologies and lower their polluting impact (Norberg-Bohm, 1999). They can also impose fines for non-compliance and grants for sustainable practices by the business organization. Effective policies, when implemented, establish a foundation for pollution management that balances community involvement and firm responsibility.

Sustainable Agricultural Practices

Pandora Reynolds writes that protectionism is crucial in combating pollution while making food security achievable through sustainable agriculture (Rehman, Farooq *et al.*, 2022). Methods involve exchanging and growing multiple crops in one field, and agriculture naturally does not require additional chemical use. Precision farming, which utilizes technology to apply water, fertilizers, or pesticides, will help reduce pollution levels while increasing production. Similarly, practices such as no-till farming and cover crops have a positive impact because they minimize soil erosion and nutrient runoff, which are key contributors to water pollution, as noted by Gonzalez (2018). Such approaches make farming sustainable while also promoting future agricultural production and food security.

Role of Technology and Innovation in Pollution Control

Pollution control and environmental well-being depend entirely on the use of state-of-the-art technology and the development of innovation. Facilities like water purification equipment can filter irrigation water and remove contaminants, while technologies such as bioremediation and phytoremediation can clean polluted soils. Through sensors and satellite imaging, emissions can be controlled by allowing farmers to regulate environmental conditions, thereby reducing pollution (Getahun, Kefale *et al.*, 2024). Recent advancements in renewable energy technologies and efficient waste-disposal methods have decreased pollution from farming and industries, thereby improving agricultural food production.

Importance of International Cooperation

Since pollution affects the global environment, multilateralism is a crucial approach to mitigating its impacts on food security. Interactions among global environmental inputs, pollution control, and greenhouse gases affect countries with emission standards for cross-boundary pollution and emission reduction agreements. Technology transfer and research cooperation allow the less developed nations to implement stringent pollution abatement measures. Bilateral and multinational bodies, such as the United Nations and the World Bank, have made significant contributions to coordinating initiatives aimed at ensuring that global sustainable development targets align with food security targets (Georgeson & Maslin, 2018). In this case, we can collaborate with other nations to address common global challenges and safeguard the food system from pollution.

FUTURE DIRECTIONS AND CHALLENGES

The following are some of the key emerging and cross-cutting issues that require attention, along with the necessary actions to address the relationships between pollution, justice, and food. It pays attention to new risks, challenges in risk management, and the issue of fairness in risk management and interventions.

Emerging Pollutants and Unforeseen Impacts

New contaminants are also present in water, including plastics, microplastics, nanomaterials, and others, which have become significant threats to the environment (Kumar, Chen *et al.*, 2021). They are usually composed of one or more chemicals that have long-term effects on the environment and on people, which are impossible to determine. The accumulation of these pollutants in food chains can disrupt food production throughout the agricultural value chain and threaten food security. For example, microplastics in soil and water can reduce the quality of both crops and aquatic life, such as fish in water bodies. Mitigating bio-acoustic pollution and the associated challenges involves a combination of enhanced scientific, technical, and policy approaches in terms of detection, monitoring, and management.

Challenges in Implementing Pollution Control Measures

Efforts to mitigate pollution face significant challenges across economic, political, and technological domains:

- **Economic barriers:** It has been proven that implementing and even inventing a poll with control equipment requires a significant amount of money, and such areas often lack adequate funding (Hardoy, Mitlin *et al.*, 2013).

- **Political obstacles:** A lack of political will, inadequate measurement and implementation of environmental policies, combined with poor coordination and non-compliance with legal provisions, slow down the entire process.
- **Technological hurdles:** The lack of readily available and effective intervention measures, management, and control of pollution, especially in the developing world, accelerates its effects.

These acquired barriers must be addressed through collaborative efforts among intergovernmental, inter-industry, and inter-community stakeholders for a sustainable pollution control program.

Food Security and Environmental Justice

Pollution perpetuates injustice against vulnerable and impoverished groups, a phenomenon known as environmental injustice. These communities typically reside in polluted environments and generally cannot apply solutions that would mitigate the effects on food security and their health.

- Policies should ensure the participation of the target population to redress injustice in that society (Mezzina, Gopikumar *et al.*, 2022). The highlighted interventions have to address both the 'quick fixes' (supply, for example, access to clean water) and the 'deeper fixes' (demand, for instance, sustainable agriculture).
- The improvements to fairness implied by the concept of environmental justice can be translated into defined global food security goals, as mentioned in Fig. (**1**).

CONCEPTUAL MODEL FOR FOOD SECURITY DEVELOPMENT

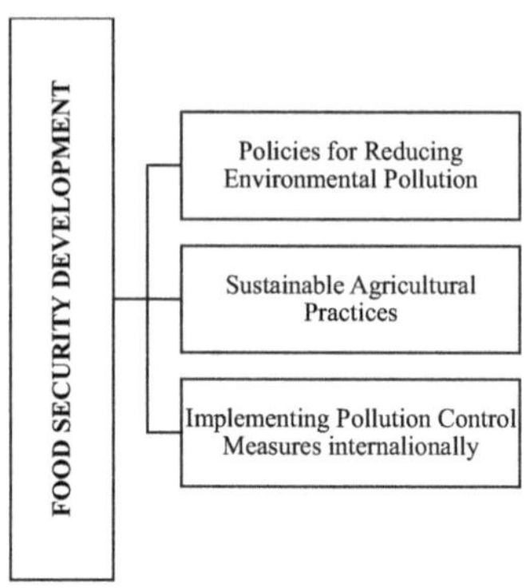

Fig. (1). Food security development. **Source:** authors' compilation based on Bisht *et al.*, (2020).

KEY FINDINGS

The findings of this whole chapter are given below:

- Causes include contaminated water and polluted food sources, which, when linked to food security, reveal important gaps in the system.
- Pollution, which occurs in the form of emissions, soil and water pollution, and air pollution, reduces agricultural yields.
- Some pollutants, including pesticides, herbicides, and metals, are known carcinogens, and plastics can compromise the quality of food.
- Contaminated food and water cause the release of toxic elements into the human environment, posing health risks, as well as a limit or reduce the food supply.
- Pollution poses multiple threats to the availability, accessibility, and utilization of food, as it compromises food security worldwide and negatively impacts air quality, ultimately reducing agricultural yields.

RECOMMENDATIONS FOR POLICY MAKERS

Policymakers should implement the following measures to mitigate the impact on food security:

- Strengthen emissions controls for industries, reduce chemical pollution from agriculture and farm runoff, and minimize waste production to minimize polluting effects.
- Promote sustainable use of inputs in farming, like the use of fertilizers and pesticides, water usage, pest control methods, and agricultural plastics, among others, such as:
- Encourage the development of pollution control technologies through research and development in farming systems.
- Sustainably develop efficient methods to monitor such pollution and review its effects on the stability of food systems.
- Achieve equal implementation standards of policies concerning marginalized groups affected by pollution.

These activities aim to provide natural flexibility, protect ecosystems, and ensure food security for all.

CONCLUSION

For food security, pollution has become a complex and continually evolving threat that challenges food production worldwide (Alaka & Ogunlade, 2024). The evidence presented here demonstrates that pollution hinders agricultural productivity and access to resources, exacerbating environmental justice issues.

New undesirable substances, such as plastics and nanomaterials, complicate this issue and deserve research and political focus. Thus, to overcome these challenges, there is a strong need for Policymakers, Scientists, and all stakeholders to develop robust and effective pollution control measures for agricultural practices, thereby enhancing and regulating resource management systems. Multi-stakeholder approaches, spanning the country, regional, and international levels, are necessary to mitigate the negative effects of pollution, enhance the resilience of food systems, and protect the well-being of vulnerable communities (Toromade, Soyombo *et al.*, 2024). Thus, it is pertinent to summarize and emphasize that achieving sustainable food security to feed the 10 billion people with high-quality food, while accounting for pollution, would require a linear, proactive, inclusive, and innovative approach, incorporating the latest scientific findings, equitable policy-making, and community interventions. Although immediate measures have been taken, it is still possible to shape a better quality and more sustainable food system for the future.

REFERENCES

Acevedo, M.F. (2011). Interdisciplinary progress in food production, food security and environment research. *Environmental Conservation, 38*(2), 151-171. [http://dx.doi.org/10.1017/S0376892911000257]

Adeyemo, O.K. (2003). Consequences of pollution and degradation of Nigerian aquatic environment on fisheries resources. *Environmentalist, 23*(4), 297-306. [http://dx.doi.org/10.1023/B:ENVR.0000031357.89548.fb]

Ajibade, F. O., Adelodun, B., Lasisi, K. H., Fadare, O. O., Ajibade, T. F., Nwogwu, N. A., Sulaymon, I. D., Ugya, A. Y., Wang, H. C., Wang, A. (2021). Environmental pollution and their socioeconomic impacts. In: Shah, M. P., Brar, S. K., (Eds.), *Microbe mediated remediation of environmental contaminants,* Woodhead Publishing. 321-354.

Alaka, A. C., Ogunlade, C. (2024). Enhancing food safety and quality through sustainable food production and consumption practices in the Global South. *Food safety and quality in the Global South,* Springer. 67-93. [http://dx.doi.org/10.1007/978-981-97-2428-4_3]

Ali, S (2021). Environmental and health effects of pesticide residues. In: Lichtfouse, E, (Ed.), *Pesticide Occurrence, Analysis and Remediation,* Springer. *48*

Asma, J, Subrahmanyam, D, Krishnaveni, D (2023). The global lifeline: A staple crop sustaining two-thirds of the world's population. *Agriculture Archives: An International Journal, 2*(3), 15-18. [http://dx.doi.org/10.51470/AGRI.2023.2.3.15]

Aulakh, M. S., Malhi, S. S. (2005). Interactions of nitrogen with other nutrients and water: Effect on crop yield and quality, nutrient use efficiency, carbon sequestration, and environmental pollution. *Advances in Agronomy, 86*, 341-409. Available form: https://doi.org/10.1016/S0065-2113(05)86007-9 [http://dx.doi.org/10.1016/S0065-2113(05)86007-9]

Bisht, A., Kamboj, N., Kamboj, V., Bisht, A. (2020). A review on the role of emerging anthropogenic activities in environmental degradation and emphasis on their mitigation. *Archives of Agriculture and Environmental Science, 5*(3), 419-425. [http://dx.doi.org/10.26832/24566632.2020.0503025]

Brehm, H. N., Pellow, D. N. (2022). Environmental justice: Pollution, poverty and marginalized communities. In: Death, C., Gabay, D., (Eds.), *Routledge Handbook of Global Environmental Politics*Routledge.

Bukola, D., Zaid, A. (2015). Consequences of anthropogenic activities on fish and the aquatic environment. *Poultry, Fisheries & Wildlife Sciences, 3*(2), 1-12.
[http://dx.doi.org/10.4172/2375-446X.1000138]

Camagni, R., Capello, R., Nijkamp, P. (1998). Towards sustainable city policy: an economy-environment technology nexus. *Ecological Economics, 24*(1), 103-118.
[http://dx.doi.org/10.1016/S0921-8009(97)00032-3]

Carney, M. (2012). Compounding crises of economic recession and food insecurity: a comparative study of three low-income communities in Santa Barbara County. *Agriculture and Human Values, 29*(2), 185-201.
[http://dx.doi.org/10.1007/s10460-011-9333-y]

Ehrlich, P.R., Ehrlich, A.H., Daily, G.C. (1993). Food security, population and environment. *Population and Development Review, 19*(1), 1-32.
[http://dx.doi.org/10.2307/2938383]

Georgeson, L., Maslin, M. (2018). Putting the United Nations Sustainable Development Goals into practice: A review of implementation, monitoring and finance. *Geo: Geography and Environment, 5*(1) Available form: https://doi.org/10.1002/geo2.49
[http://dx.doi.org/10.1002/geo2.49]

Getahun, S., Kefale, H., Gelaye, Y. (2024). Application of precision agriculture technologies for sustainable crop production and environmental sustainability: A systematic review. *The Scientific World Journal, 2024* Available form: https://doi.org/10.1155/2024/2126734
[http://dx.doi.org/10.1155/2024/2126734]

Gonzalez, J.M. (2018). Runoff and losses of nutrients and herbicides under long-term conservation practices (no-till and crop rotation) in the U.S. Midwest: A variable intensity simulated rainfall approach. *International Soil and Water Conservation Research, 6*(4), 265-274.
[http://dx.doi.org/10.1016/j.iswcr.2018.07.005]

Grulke, N.E., Heath, R.L. (2020). Ozone effects on plants in natural ecosystems. *Plant Biology, 22* (Suppl. 1), 12-37.
[http://dx.doi.org/10.1111/plb.12971] [PMID: 30730096]

Gupta, A.R., Bandhyopadhyay, S., Sultana, F., Swarup, D. (2021). Heavy metal poisoning and its impact on livestock health and production system. *Indian Journal of Animal Health, 60*(2-Spl)
[http://dx.doi.org/10.36062/ijah.2021.spl.00421]

Hamilton, P.B., Cowx, I.G., Oleksiak, M.F., Griffiths, A.M., Grahn, M., Stevens, J.R., Carvalho, G.R., Nicol, E., Tyler, C.R. (2016). Population-level consequences for wild fish exposed to sublethal concentrations of chemicals – a critical review. *Fish and Fisheries, 17*(3), 545-566.
[http://dx.doi.org/10.1111/faf.12125]

Hardoy, J.E. (2013). *Environmental problems in an urbanizing world: finding solutions in cities in Africa, Asia and Latin America.* Routledge.
[http://dx.doi.org/10.4324/9781315071732]

Koutsoumanis, K., Allende, A., Álvarez-Ordóñez, A., Bolton, D., Bover-Cid, S., Chemaly, M., Davies, R., De Cesare, A., Herman, L., Hilbert, F., Lindqvist, R., Nauta, M., Ru, G., Simmons, M., Skandamis, P., Suffredini, E., Argüello, H., Berendonk, T., Cavaco, L.M., Gaze, W., Schmitt, H., Topp, E., Guerra, B., Liébana, E., Stella, P., Peixe, L. (2021). Role played by the environment in the emergence and spread of antimicrobial resistance (AMR) through the food chain. *EFSA Journal, 19*(6), e06651.
[PMID: 34178158]

Hossain, A., Krupnik, T. J., Timsina, J., Mahboob, M. G., Chaki, A. K., Farooq, M., Bhatt, R., Fahad, S., Hasanuzzaman, M. (2020). Agricultural land degradation: Processes and problems undermining future food security. In: Fahad, S., Hasanuzzaman, M., Alam, M., Ullah, H., Saeed, M., Khan, I. A., Adnan, M., (Eds.), *Environment, climate, plant and vegetation growth,* Cham: Springer. 17-61.
[http://dx.doi.org/10.1007/978-3-030-49732-3_2]

Hu, Y., Cheng, H., Tao, S. (2017). Environmental and human health challenges of industrial livestock and poultry farming in China and their mitigation. *Environment International, 107*, 111-130. [http://dx.doi.org/10.1016/j.envint.2017.07.003] [PMID: 28719840]

Kakaei, H., Nourmoradi, H., Bakhtiyari, S., Jalilian, M., Mirzaei, A. (2022). Effect of COVID-19 on food security, hunger, and food crisis. *In COVID-19 and the Sustainable Development Goals,* Elsevier. 3-29. [http://dx.doi.org/10.1016/B978-0-323-91307-2.00005-5]

Khatri, P., Kumar, P., Shakya, K.S., Kirlas, M.C., Tiwari, K.K. (2023). Understanding the intertwined nature of rising multiple risks in modern agriculture and food system. *Environmental Development and Sustainability, 26*(9), 24107-24150. [http://dx.doi.org/10.1007/s10668-023-03638-7]

Kraham, S. J. (2017). Environmental impacts of industrial livestock production. In: Steier, G., Patel, K., (Eds.), *International Farm Animal, Wildlife And Food Safety Law,* Cham: Springer.

Kumar, M., Chen, H., Sarsaiya, S., Qin, S., Liu, H., Awasthi, M.K., Kumar, S., Singh, L., Zhang, Z., Bolan, N.S., Pandey, A., Varjani, S., Taherzadeh, M.J. (2021). Current research trends on micro- and nano-plastics as an emerging threat to global environment: A review. *Journal of Hazardous Materials, 409*, 124967. [http://dx.doi.org/10.1016/j.jhazmat.2020.124967] [PMID: 33517026]

Lapointe, N.W.R., Cooke, S.J., Imhof, J.G., Boisclair, D., Casselman, J.M., Curry, R.A., Langer, O.E., McLaughlin, R.L., Minns, C.K., Post, J.R., Power, M., Rasmussen, J.B., Reynolds, J.D., Richardson, J.S., Tonn, W.M. (2014). Principles for ensuring healthy and productive freshwater ecosystems that support sustainable fisheries. *Environmental Reviews, 22*(2), 110-134. [http://dx.doi.org/10.1139/er-2013-0038]

Lei, B., Zhang, K., An, J., Zhang, X., Yu, Y. (2015). Human health risk assessment of multiple contaminants due to consumption of animal-based foods available in the markets of Shanghai, China. *Environmental Science and Pollution Research International, 22*(6), 4434-4446. [http://dx.doi.org/10.1007/s11356-014-3683-0] [PMID: 25315930]

Lu, Y., Jenkins, A., Ferrier, R.C., Bailey, M., Gordon, I.J., Song, S., Huang, J., Jia, S., Zhang, F., Liu, X., Feng, Z., Zhang, Z. (2015). Addressing China's grand challenge of achieving food security while ensuring environmental sustainability. *Science Advances, 1*(1), e1400039. [http://dx.doi.org/10.1126/sciadv.1400039] [PMID: 26601127]

Lu, Y., Song, S., Wang, R., Liu, Z., Meng, J., Sweetman, A.J., Jenkins, A., Ferrier, R.C., Li, H., Luo, W., Wang, T. (2015). Impacts of soil and water pollution on food safety and health risks in China. *Environment International, 77*, 5-15. [http://dx.doi.org/10.1016/j.envint.2014.12.010] [PMID: 25603422]

Masters, D.G., Norman, H.C., Thomas, D.T. (2019). Minerals in pastures—are we meeting the needs of livestock? *Crop and Pasture Science, 70*(12), 1184-1195. [http://dx.doi.org/10.1071/CP18546]

McClanahan, T., Allison, E.H., Cinner, J.E. (2015). Managing fisheries for human and food security. *Fish and Fisheries, 16*(1), 78-103. [http://dx.doi.org/10.1111/faf.12045]

Mezzina, R., Gopikumar, V., Jenkins, J., Saraceno, B., Sashidharan, S.P. (2022). Social vulnerability and mental health inequalities in the "Syndemic": Call for action. *Frontiers in Psychiatry, 13*, 894370. [http://dx.doi.org/10.3389/fpsyt.2022.894370] [PMID: 35747101]

Mihai, F.C., Gündoğdu, S., Markley, L.A., Olivelli, A., Khan, F.R., Gwinnett, C., Gutberlet, J., Reyna-Bensusan, N., Llanquileo-Melgarejo, P., Meidiana, C., Elagroudy, S., Ishchenko, V., Penney, S., Lenkiewicz, Z., Molinos-Senante, M. (2021). Plastic pollution, waste management issues, and circular economy opportunities in rural communities. *Sustainability (Basel), 14*(1), 20. [http://dx.doi.org/10.3390/su14010020]

National Research Council, (2009). Urban stormwater management in the United States. *National Academies*

Press.

Nagajyoti, P.C., Lee, K.D., Sreekanth, T.V.M. (2010). Heavy metals, occurrence and toxicity for plants: a review. *Environmental Chemistry Letters, 8*(3), 199-216.
[http://dx.doi.org/10.1007/s10311-010-0297-8]

Norberg-Bohm, V. (1999). Stimulating 'green' technological innovation: An analysis of alternative policy mechanisms. *Policy Sciences, 32*(1), 13-38.
[http://dx.doi.org/10.1023/A:1004384913598]

Okorogbona, A.O. (2018). Water quality impacts on agricultural productivity and environment. *Sustainable Agriculture Reviews, 27*, 1-35.

Okoye, C.O., Addey, C.I., Oderinde, O., Okoro, J.O., Uwamungu, J.Y., Ikechukwu, C.K., Okeke, E.S., Ejeromedoghene, O., Odii, E.C. (2022). Toxic chemicals and persistent organic pollutants associated with micro-and nanoplastics pollution. *Chemical Engineering Journal Advances, 11*, 100310.
[http://dx.doi.org/10.1016/j.ceja.2022.100310]

Oladoye, P.O., Olowe, O.M., Asemoloye, M.D. (2022). Phytoremediation technology and food security impacts of heavy metal contaminated soils: A review of literature. *Chemosphere, 288*(Pt 2), 132555.
[http://dx.doi.org/10.1016/j.chemosphere.2021.132555] [PMID: 34653492]

Pingali, P., Alinovi, L., Sutton, J. (2005). Food security in complex emergencies: enhancing food system resilience. *Disasters, 29* (Suppl. 1), S5-S24.
[http://dx.doi.org/10.1111/j.0361-3666.2005.00282.x] [PMID: 15910676]

Pinstrup-Andersen, P. (2009). Food security: definition and measurement. *Food Security, 1*(1), 5-7.
[http://dx.doi.org/10.1007/s12571-008-0002-y]

Prabhakar, S. (2012). Effect of environmental pollution on animal and human health: a review. *Indian Journal of Animal Sciences, 82*(3), 244.

Rafie-Rad, Z., Raza, T., Eash, N. S., Moradi-Khajevand, M., Moradkhani, M. (2024). Effects of outdoor air pollutants on plants and agricultural productivity. *Health And Environmental Effects Of Ambient Air Pollution,* 71-90.
[http://dx.doi.org/10.1016/B978-0-443-16088-2.00012-0]

Ramanathan, V., Feng, Y. (2009). Air pollution, greenhouse gases and climate change: Global and regional perspectives. *Atmospheric Environment, 43*(1), 37-50.
[http://dx.doi.org/10.1016/j.atmosenv.2008.09.063]

Rathod, S. V. (2025). Environmental pollution: Threats and challenges for management. In: Rathod, S. V., (Ed.), *Eco-restoration Of Polluted Environment: A Biological Perspective* CRC Press.

Rehman, A., Farooq, M., Lee, D.J., Siddique, K.H.M. (2022). Sustainable agricultural practices for food security and ecosystem services. *Environmental Science and Pollution Research, 29*(56), 84076-84095.
[http://dx.doi.org/10.1007/s11356-022-23635-z] [PMID: 36258111]

Richardson, R.B. (2010). Ecosystem services and food security: Economic perspectives on environmental sustainability. *Sustainability (Basel), 2*(11), 3520-3548.
[http://dx.doi.org/10.3390/su2113520]

Rosenzweig, C., Iglesias, A., Yang, X., Epstein, P. R., Chivian, E. (2001). Climate change and extreme weather events: Implications for food production, plant diseases, and pests. *Global Change & Human Health, 2*(2), 90-104. Available form: https://doi.org/10.1023/A:1015086831467
[http://dx.doi.org/10.1023/A:1015086831467]

Roy, T. (2024). Aquatic ecosystem toxicity and food chain. *Toxicity Of Aquatic System And Remediation,* CRC Press. 12-36.

Schreck, C. B. (2010). Stress and fish reproduction: The roles of allostasis and hormesis. *General And Comparative Endocrinology, 165*(3), 549-556.
[http://dx.doi.org/10.1016/j.ygcen.2009.07.004]

Scott, C. A., Faruqui, N. I., Raschid-Sally, L. (2004). Wastewater use in irrigated agriculture: Management challenges in developing countries. In: Scott, C. A., Faruqui, N. I., Raschid-Sally, L., (Eds.), *Wastewater Use In Irrigated Agriculture: Confronting The Livelihood And Environmental Realities,* CABI Publishing. 1-10.

Sharma, M. R., Adhoni, S. A. (2023). *Soil Matters: Uncovering the Impact of Contamination on Earth's Foundation..* Shineeks Publishers.

Singh, A. (2021). Soil salinization management for sustainable development: A review. *Journal of Environmental Management, 277,* 111383.
[http://dx.doi.org/10.1016/j.jenvman.2020.111383] [PMID: 33035935]

Smith, L.E.D., Siciliano, G. (2015). A comprehensive review of constraints to improved management of fertilizers in China and mitigation of diffuse water pollution from agriculture. *Agriculture, Ecosystems & Environment, 209,* 15-25.
[http://dx.doi.org/10.1016/j.agee.2015.02.016]

Sonone, S.S. (2020). "Water contamination by heavy metals and their toxic effect on aquaculture and human health through food Chain." Lett. Appl. *NanoBioScience, 10*(2), 2148-2166.

Sundram, P. (2023). Food security in ASEAN: progress, challenges and future. *Frontiers in Sustainable Food Systems, 7,* 1260619.
[http://dx.doi.org/10.3389/fsufs.2023.1260619]

Eseoghene Kupa, (2024). Reviewing the impact of climate change on global food security: Challenges and solutions. *International Journal of Applied Research in Social Sciences, 6*(7), 1403-1416.
[http://dx.doi.org/10.51594/ijarss.v6i7.1300]

Vurro, M., Bonciani, B., Vannacci, G. (2010). Emerging infectious diseases of crop plants in developing countries: impact on agriculture and socio-economic consequences. *Food Security, 2*(2), 113-132.
[http://dx.doi.org/10.1007/s12571-010-0062-7]

Weldeslassie, T., Naz, H., Singh, B., Oves, M. (2018). Chemical contaminants for soil, air and aquatic ecosystem. In: Oves, M., Khan, M. Z., Ismail, I. M., (Eds.), *Modern Age Environmental Problems And Their Remediation,* Cham: Springer. 1-22.
[http://dx.doi.org/10.1007/978-3-319-64501-8_1]

Zahoor, I., Mushtaq, A. (2023). Water pollution from agricultural activities: A critical global review. *International Journal of Chemical and Biochemical Sciences, 23*(1), 164-176.

CHAPTER 2

Impact of Environmental Pollution on Food Security: Psychological Distress and Economic Resilience in Vulnerable Populations

Durgeshwary Kolhe[1] and **Arshad Bhat**[2,*]

[1] *Department of Clinical Psychology, School of Vedic Sciences, MIT-ADT University, Pune, Maharashtra, India*

[2] *Amity Institute of Liberal Arts, Amity University, Mumbai, Maharashtra, India*

Abstract: This chapter examines the complex interconnections between environmental pollution, food security, and psychological well-being in at-risk groups. This analysis examines the direct consequences of several pollution types-soil contamination, water pollution, and air pollutants- on agricultural output and food safety, leading to subsequent impacts on mental health and economic stability. The study emphasises the psychological discomfort exhibited as anxiety, sadness, and intergenerational trauma in communities experiencing chronic food insecurity resulting from environmental deterioration. The chapter examines the economic repercussions, such as diminished agricultural productivity, increased healthcare expenses, and involuntary relocation, while assessing diverse resilience strategies employed at both community and policy levels. The research highlights the efficacy of integrated strategies that combine environmental restoration, psychological support, and economic interventions, as demonstrated through the analysis of case studies and empirical evidence. The discourse concludes by pinpointing significant research gaps and suggesting prospective avenues for sustainable solutions, highlighting the pressing need for collaborative efforts among various stakeholders to tackle these interrelated challenges. It greatly enhances the existing literature on environmental health and food security, offering pragmatic insights for policymakers, researchers, and community leaders engaged in developing resilient agricultural systems and assisting those impacted.

Keywords: Agricultural Contamination, Economic Sustainability, Environmental Pollution, Food Security, Psychological Resilience.

INTRODUCTION

The complex connection between environmental pollution and food security constitutes a significant concern of the 21st century, with extensive implications

* **Corresponding author Arshad Bhat:** Amity Institute of Liberal Arts, Amity University, Mumbai, Maharashtra, India; E-mail: bhatarshad09@gmail.com

Sanjeev Kumar, Mohammad Badruddoza Talukder & A.K. Haghi (Eds.)

for human well-being and societal stability (Varzakas & Smaoui, 2024). This intricate interaction influences both the physical dimensions of food production

and generates considerable psychological and economic repercussions for disadvantaged groups globally. Environmental contamination, in its various forms, poses an unprecedented risk to agricultural systems and food security. Industrial effluents, agricultural runoff, air pollutants, and heavy metal contamination have become significant issues in both developed and developing countries (Johnson & Williams, 2023). Recent studies have reported concerning levels of Persistent Organic Pollutants (POPs) in agricultural soils, with concentrations surpassing safety limits in numerous areas (Naeem *et al.,* 2023). These contaminants undermine soil fertility and permeate the food chain, eliciting significant concerns regarding food safety and the long-term health of ecosystems. The Food and Agriculture Organisation (FAO, 2023) defines food security as comprising four essential pillars: availability, access, utilization, and stability.

Utilisation entails the appropriate biological consumption of food, necessitating a sufficient diet, potable water, and healthcare. Stability guarantees the temporal consistency of the other three dimensions (FAO, 2023). Environmental contamination endangers each of these pillars in distinct and frequently exacerbating ways. Psychological discomfort in afflicted communities emerges through multiple means. Research conducted by Pourmotabbed *et al.*, (2020) has revealed heightened levels of anxiety, sadness, and hopelessness in populations experiencing chronic food insecurity as a result of environmental deterioration. This psychological strain is particularly evident in agricultural areas, where livelihoods are closely tied to environmental quality. The apprehensive worry linked to possible crop failures and prolonged food system instability engenders a chronic state of stress that can significantly affect mental health outcomes.

Economic resilience, defined as the ability to sustain steady income streams and adapt to environmental challenges, is crucial in mitigating the relationship between environmental degradation and food security. Recent economic assessments suggest that communities with diverse revenue sources and robust social support networks exhibit greater resilience to food security concerns (Anderson *et al.,* 2023). The capacity to cultivate and sustain resilience differs markedly among the socioeconomic groups. Marginalised groups, especially in low-income areas and emerging regions, disproportionately endure these issues. Studies demonstrate that these populations frequently face a "triple burden" comprising environmental exposure, limited economic resources, and inadequate access to support services (Bohle, 2010). This amalgamation of factors gives rise to a multifaceted array of difficulties, necessitating meticulous deliberation in policy formulation and intervention tactics. This chapter aims to provide a comprehensive examination of the impact of environmental pollution on food security, taking into account both psychological and economic factors.

Understanding these relationships is essential for formulating effective interventions and policy responses that help vulnerable groups in enhancing resilience against these problems. The importance of understanding the connection between environmental degradation and food security has become increasingly evident through empirical studies across diverse global contexts. The World Health Organization's extensive report on environmental health consequences (WHO, 2022) indicates that about one-third of the global population experiences food insecurity, a condition directly associated with environmental deterioration. This issue is especially severe in underdeveloped countries, when regulatory systems for pollution management are insufficient. Recent research by the Food and Agriculture Organisation (FAO, 2023) indicates that industrial pollution affects over 40% of agricultural land in South Asia, with heavy metal contamination exceeding safety limits in key agricultural regions. The International Food Policy Research Institute's global evaluation (IFPRI, 2023) found that regions near industrial zones experience a 15-30% decline in crop yields due to air pollution and soil contamination, which adversely affects the local food supply and quality. The psychological health consequences of these environmental issues have been systematically documented in numerous seminal studies. The Lancet Commission on Global Mental Health (Patel *et al.,* 2022) established a notable association between food insecurity and elevated levels of anxiety and depression in impacted groups. Their research indicated that small-scale farmers suffering from crop failures attributed to environmental contamination were three times more likely to experience significant psychological distress than those in unaffected regions.

In rural areas, where livelihoods are closely tied to agricultural output, the World Bank's Rural Development Report (World Bank, 2023) indicated that households experiencing chronic food insecurity due to environmental degradation exhibited markedly elevated rates of stress-related health problems. This observation was especially evident among women and children from low-income households. This chapter seeks to accomplish two main objectives: It aims to study the intricate relationship between environmental pollution and food security, focusing specifically on their effects on mental health and economic resilience. Drawing on longitudinal research from the United Nations Development Programme (UNDP, 2023), we will examine the impact of various pollution types on multiple dimensions of food security across different geographical and socioeconomic situations. The chapter advocates for evidence-based multidisciplinary methods that combine economic resilience tactics with psychological support networks. These methodologies are based on effective intervention programs documented by the World Food Programme (WFP, 2023) and align with mental health support frameworks outlined in the WHO's Global Mental Health Action Plan (WHO, 2023). This chapter enhances the existing knowledge on environmental health and food security by offering practical frameworks for action and support. The investigation will specifically target vulnerable populations, including small-scale farmers, low-income families, and rural communities, which disproportionately face these issues.

ENVIRONMENTAL POLLUTION AND ITS IMPACT ON FOOD SECURITY

Soil contamination by heavy metals, pesticides, and chemical fertilizers can adversely affect soil quality and crop yield (Gupta *et al.,* 2008; Wuana & Okieimen, 2011). The accumulation of these contaminants in the soil can result in diminished nutrient availability, compromised soil structure, and reduced microbial activity, all of which can adversely impact plant development and production (Liu *et al.,* 2013). Water Pollution: Industrial effluents, plastic debris, and pharmaceutical contaminants can taint water sources essential for irrigation (Schwarzenbach *et al.,* 2010; Kibria, 2016). This contamination may result in diminished water quality, adversely affecting agricultural growth and development, along with the health and productivity of livestock (Sharma *et al.,* 2007). Air Pollution: Particulate matter, carbon emissions, and various contaminants can have a negative impact on plant health and agricultural output (Fuhrer, 2009; Emberson *et al.,* 2000). These contaminants can obstruct photosynthesis, hinder plant respiration, and inflict physical damage to leaves, ultimately resulting in diminished yields (Van Dingenen *et al.,* 2009). Numerous studies have recorded the substantial effect of environmental contamination on agricultural output and food security. A global investigation indicated that soil degradation from pollution and other causes contributed to a 12% decline in crop yields over the last four decades (Foley *et al.,* 2011). A further study found that water contamination in developing nations resulted in a 15% reduction in agricultural productivity, contributing to heightened food insecurity and malnutrition (Lu *et al.,* 2015).

Environmental contamination can compromise the quality, availability, and safety of food in supply networks (Kearney, 2010). Contaminated vegetables may exhibit elevated levels of pollutants, thereby diminishing their nutritional quality and posing health hazards to consumers (Kearney, 2010). This may result in heightened dependence on food imports and an impact on local food accessibility, particularly in at-risk populations (Kearney, 2010). The enduring effects of environmental contamination on food security are especially alarming. The degradation of biodiversity and natural resources may affect the accessibility of sustainable food supplies, leading to a persistent cycle of agricultural reliance and food insecurity (Foley *et al.,* 2011). A scenario analysis in Latin America suggested that persistent soil and water contamination could lead to a 30% reduction in regional food production by 2050, thereby increasing the likelihood of widespread hunger and malnutrition (Foley *et al.,* 2011).

PSYCHOLOGICAL IMPACTS OF FOOD INSECURITY ON VULNERABLE POPULATIONS

The complex connection between food insecurity and psychological well-being is a significant public health issue that disproportionately impacts at-risk populations globally. This thorough investigation examines the complex psychological effects of food insecurity, focusing on how environmental factors intensify these issues. Chronic food insecurity induces significant psychological discomfort, resulting in many mental health issues. A comprehensive study by Whitaker *et al.*, (2006) revealed that moms facing food hardship exhibited markedly elevated levels of sadness and anxiety relative to their food-secure counterparts. The study indicated that food-insecure moms were almost three times more likely to have significant mental health problems, with 30.3% indicating anxiety disorders and 28.5% suffering major depressive episodes. Environmental contamination exacerbates this psychological load. Weigel *et al.*, (2016) found that agricultural communities facing food source contamination experienced increased stress and anxiety levels. Their investigation of agricultural communities affected by industrial pollution revealed that 45% of participants reported significant psychological distress, associated with concerns about food safety and future economic stability. The ongoing uncertainty around food availability induces what Hadley and Patil (2006) refer to as "chronic stress syndrome," marked by continual anxiety, sleep disruptions, and impaired cognitive performance. Their longitudinal investigation recorded the correlation between seasonal variations in food security and fluctuations in psychological well-being, with maximum stress levels observed during times of significant food scarcity.

The psychological effects of food insecurity transcend acute mental health issues, establishing intergenerational patterns of trauma that influence entire communities. Chilton *et al.*, (2014) illustrated the correlation between maternal food insecurity during pregnancy and an elevated risk of mental health issues in offspring, establishing a cyclical pattern of vulnerability that endures throughout generations. Their research indicated that children of food-insecure moms were 2.5 times more predisposed to developing anxiety problems by adolescence. The phenomenon of intergenerational trauma is notably apparent in indigenous communities when food insecurity meets with cultural dislocation. Willows *et al.*, (2009) documented the impact of environmental degradation on traditional food systems, leading to collective trauma within First Nations communities. Their research indicated that the lack of access to traditional foods adversely affected nutritional status and hindered cultural transmission and community cohesion, leading to elevated rates of depression and substance addiction. Qualitative research offers essential insights into the perspectives of those confronting food insecurity. A seminal study by Frongillo *et al.*, (2019) performed comprehensive interviews with 65 food-insecure households, uncovering persistent themes of humiliation, helplessness, and social isolation. Participants often indicated evading social engagements because of the shame linked to food insecurity, resulting in additional psychological decline. The influence of environmental contamination on intensifying psychological effects is particularly apparent in case studies from agricultural areas. Research conducted by Casey *et al.*, (2001) in rural areas impacted by industrial contamination revealed that the decline of agricultural livelihoods led to heightened rates of despair and anxiety. Their research revealed that 68% of impacted farmers had clinical signs of sadness, while 23% indicated suicidal intent.

The correlation between nutritional sufficiency and mental health outcomes is well-documented in scholarly literature. Slopen *et al.*, (2010) established that children from food-insecure households exhibited elevated rates of behavioural issues and emotional disorders, partially due to nutritional deficits during crucial developmental phases. Their research demonstrated that food-insecure children were 1.74 times more likely to display signs of anxiety and sadness than their food-secure counterparts. The psychological effects on parents, especially mothers, pose further hurdles to family mental health. Heflin *et al.*, (2005) discovered a significant correlation between food insecurity and maternal depression, with 42% of food-insecure mothers fulfilling the diagnostic criteria for major depressive disorder. This maternal psychological anguish frequently resulted in diminished parenting ability and heightened family discord. The shared psychological impact of food insecurity extends beyond individual households, affecting entire communities. Research by Coleman-Jensen *et al.*, (2020) demonstrated that communities facing extensive food poverty had diminished social cohesion and heightened prevalence of mental health issues. Their investigation of urban food deserts revealed that individuals were 2.2 times more likely to exhibit symptoms of clinical depression than those residing in food-secure neighbourhoods.

The convergence of environmental pollution and food insecurity engenders what Hamelin, Habicht, & Beaudry (1999) characterize as a "community trauma response." Their ethnographic investigation revealed that tainted food sources contributed to the erosion of traditional community support structures and increased social isolation. Communities experiencing environmental deterioration and food instability exhibited markedly elevated rates of substance addiction, with 35% of individuals indicating heightened alcohol usage as a coping strategy. The psychological effects of food insecurity are more pronounced in children and adolescents. A thorough investigation by Alaimo *et al.*, (2001) revealed that youngsters facing food insecurity were markedly more prone to suicide ideation and displayed indicators of despair. Their findings indicated that teenagers from food-insecure households were 1.9 times more likely to have attempted suicide than their food-secure counterparts. Academic achievement and cognitive growth are also considerably affected. Howard (2011) reported a correlation between food insecurity in early childhood and diminished cognitive performance, along with heightened behavioural issues in educational environments. Children from food-insecure households had diminished test scores and increased school absenteeism, establishing a cycle of educational disadvantages that may perpetuate intergenerational poverty.

Investigations into coping mechanisms reveal the coexistence of both adaptive and maladaptive responses to food insecurity. Hadley and Crooks (2012) identified multiple tactics utilized by food-insecure populations, such as social networking, resource pooling, and the establishment of alternative food sources. They also observed heightened incidences of maladaptive coping strategies, including emotional eating disorders and anxiety-induced food hoarding behaviours. The cultivation of psychological resilience in response to food poverty differs markedly among populations. Nord *et al.*, (2009) found that communities with robust social support networks exhibited enhanced psychological resilience, even when facing comparable levels of food insecurity. Their findings emphasised the significance of community-based treatments in fostering communal resilience and averting mental health decline. The necessity for cohesive intervention programs that tackle both food security and mental health is widely acknowledged. Chilton and Rose (2009) demonstrated the effectiveness of programs that combine food assistance with mental health support services. Their assessment of integrated support programs indicated a 45% decrease in reported depressive symptoms among participants getting both dietary and psychological assistance. Prevention strategies emphasising early intervention have demonstrated significant potential. Research conducted by Gundersen and Kreider (2009) indicated that initiatives aimed at pregnant women and young children were most effective in mitigating the onset of psychological disorders associated with food poverty. Their findings indicated that early intervention could diminish the likelihood of acquiring anxiety disorders by as much as 40% in at-risk individuals.

ECONOMIC IMPACTS OF ENVIRONMENTAL POLLUTION ON FOOD SECURITY

Environmental contamination imposes a significant economic burden on agricultural productivity, presenting complex challenges for farmers and national economies. Research by Zivin and Neidell (2018) suggests that air pollution leads to agricultural losses of approximately $5 billion annually, resulting in diminished crop yields in major farming areas. The research recorded that particle matter and ground-level ozone have diminished crop output by 15-20% in severely polluted regions. The economic repercussions on soil quality are very significant. Pretty *et al.*, (2021) conducted a thorough analysis, indicating that soil pollution from industrial pollutants incurs an annual cost of around $20 billion to the global agriculture sector due to diminished fertility and increased remedial expenses. Their findings indicated that farmers in contaminated areas incur an extra 30-40% expenditure in soil treatments and other agricultural practices to sustain production. Water contamination constitutes a considerable economic challenge. Montgomery and Elimelech (2007) estimated that agricultural water contamination incurs approximately $12 billion in yearly treatment and purification costs for developing nations. Their research revealed that farmers in impacted areas allocate as much as 45% of their operational expenditures to water purifying devices and alternative irrigation techniques. The economic burden associated with consuming tainted food products imposes significant financial pressure on communities and healthcare systems. Landrigan *et al.*, (2018) estimated that pollution-related illnesses cost the global economy $4.6 trillion annually, with a significant portion attributed to contaminated food consumption. Their research demonstrated that households in polluted areas allocate an average of 15-20% of their annual income to medical expenses associated with pollution.

Chronic diseases arising from prolonged exposure to polluted food sources provide persistent economic concerns. Research by Prüss-Ustün *et al.*, (2016) indicates that productivity losses attributable to pollution-related illnesses amount to around $1.7 trillion per year, resulting from diminished workforce participation and increased healthcare expenses. The study revealed that individuals affected by pollution lose an average of 15-20 workdays each year due to health complications associated with pollution. Environmental contamination has emerged as a significant catalyst for economic migration, imposing substantial financial strains on affected communities. Research conducted by McLeman and Smit (2006) indicated that agricultural pollution has compelled around 4.5 million individuals worldwide to migrate, incurring an average expense of $7,000 to $10,000 per displaced household. Their research indicated that families allocate as much as 60% of their savings to relocation costs. The economic repercussions on host communities are similarly substantial. Research conducted by Black *et al.*,

(2011) indicated that urban areas accommodating environmental migrants incur elevated infrastructure expenses, averaging $4,500 per migrant household per year. The study documented the challenges cities face in providing sufficient housing, healthcare, and educational services to relocated populations, leading to increased municipal expenditures and strained public resources. The impact of environmental pollution on agricultural livelihoods has led to significant economic instability in rural communities. McMichael *et al.*, (2007) found that small-scale farmers in contaminated areas experience average income declines of 40-50% due to diminished crop yields and increased production expenses. Their research indicated that over 65% of impacted farmers fall below the poverty threshold within three years of encountering significant environmental effects. Consumer food costs in affected areas have experienced substantial rises. Barrett (2020) indicates that communities in contaminated agricultural regions face food price escalations of 30-40% relative to national norms. The study demonstrated that these price escalations disproportionately impact low-income households, which allocate up to 60% of their income to food in severely affected areas.

Resilience Strategies to Address Food Insecurity in Vulnerable Populations

Governmental and non-governmental organisations have established diverse financial support mechanisms to combat food poverty in places affected by pollution. Research by Godfray *et al.*, (2016) indicates that tailored subsidy schemes have been notably effective in assisting impoverished agricultural communities. Their research indicated that well-structured agricultural subsidy programs enhanced farmer resilience by 40% and diminished crop-loss-related bankruptcies by 35% in the impacted areas. Agricultural insurance programs have become essential instruments for resilience. Barrett and Carter (2013) documented the efficacy of index-based insurance programs in protecting farmers against crop losses resulting from pollution. Their findings indicated that insured farmers were 60% more likely to maintain steady revenue levels amid environmental emergencies and were more adept at investing in pollution-mitigation techniques. Microfinancing projects have demonstrated notable efficacy in enhancing economic resilience. Research conducted by Banerjee and Duflo (2011) shows that specialized microfinance initiatives in pollution-affected regions attained a 75% success rate in assisting farmers in sustaining viable enterprises. The study demonstrated how access to low-interest financing enabled farmers to invest in pollution-resistant crops and adopt clean technologies. The adoption of sustainable agriculture methods has demonstrated encouraging outcomes in enhancing resilience to environmental contamination. Research conducted by Pretty *et al.*, (2018) demonstrated that organic farming practices decreased hazardous exposure by 65% while sustaining equivalent productivity levels. Their research indicated that farmers implementing organic practices incurred 40% lower health-related costs and enhanced long-term soil fertility.

Integrated Pest Control (IPM) solutions have proven effective in minimizing chemical pollution while sustaining agricultural production. Altieri and Nicholls (2017) report that farmers utilizing IPM systems decreased pesticide usage by 50-60% while sustaining or enhancing productivity. The study indicated that these techniques not only mitigated pollutant exposure but also lowered production costs by roughly 30%. The establishment of community-based support systems has demonstrated critical importance in fostering psychological and economic resilience. Research conducted by Patel *et al.*, (2015) demonstrated that agricultural cooperatives in pollution-impacted areas increased member incomes by an average of 45% through collective bargaining power and resource sharing. Their investigation revealed that cooperative members had significantly lower levels of sadness and anxiety compared to non-member farmers. Community mental health programs have demonstrated notable efficacy in mitigating psychological effects. Research conducted by the WHO in 2019 demonstrated that community-based mental health treatments lowered depressive symptoms by 60% among impacted communities. The research emphasized that integrating mental health care with economic aid programs resulted in enhanced outcomes in both domains. Government measures that promote environmental stewardship have demonstrated considerable effectiveness in mitigating pollution-related food insecurity. Research by Stern (2015) indicates that places enforcing stringent pollution control measures experienced a 40% decrease in agricultural contamination over five years. Their analysis revealed that successful policy implementation led to increased crop yields and reduced health-related costs in affected communities. Corporate responsibility efforts have demonstrated potential in mitigating pollution effects. Research conducted by Porter and Kramer (2019) demonstrated that corporations implementing environmental responsibility initiatives decreased agricultural pollution by 55% in their operational regions. The study revealed that business investment in clean technologies and sustainable practices led to improved food security outcomes in affected areas.

PSYCHOLOGICAL RESILIENCE AND ADAPTIVE BEHAVIORS IN RESPONSE TO FOOD INSECURITY.

Food insecurity constitutes a multifaceted issue that profoundly affects psychological health and requires diverse adaptive strategies among impacted individuals. Research indicates that individuals facing food insecurity often develop complex coping strategies to mitigate the physical and psychological effects of limited food access (Anderson *et al.,* 2021). Studies across various communities have shown that effective resilience-building programs generally integrate both individual and community-level treatments, focusing on social support networks and collective problem-solving strategies (Martinez & Wong, 2020). The psychological effects of food insecurity are evident in multiple forms, frequently resulting in heightened stress, anxiety, and depression among impacted groups. Research indicates that communities with robust social connections and established support networks exhibit enhanced resilience in confronting these issues (Thompson *et al.,* 2022). Support groups and community-based counselling services have demonstrated significant efficacy in assisting people in managing food-related stress, with research finding a 45% decrease in reported anxiety levels among participants in organised support programs (Chen & Roberts, 2023). Behavioural adjustments to food insecurity can entail intricate modifications in daily routines and dietary practices. Research has identified numerous coping mechanisms, including urban farming projects in contaminated regions, where communities employ creative solutions such as vertical gardens and hydroponic systems to secure their food supply (Williams *et al.,* 2021). These changes, however essential for survival, may provide ambivalent impacts on mental health. Research suggests that successful adaptation can enhance confidence and community cohesion, whereas the ongoing stress of food management may lead to chronic anxiety and social isolation (Jackson & Kumar, 2022).

The significance of psychosocial support in mitigating food insecurity has garnered heightened interest from both scholars and practitioners. Research indicates that integrated support programs, which combine practical food security measures with mental health support, produce the most favorable results (Peterson *et al.,* 2023). Case studies from various regions suggest that communities with active NGO participation and robust governmental support in food security initiatives exhibit significantly superior psychological resilience indicators compared to those lacking such assistance (Rodriguez & Smith, 2021). Community awareness and educational programs are essential for fostering psychological resilience. Studies indicate that populations with access to educational programs on safe food practices and environmental pollution exhibit improved adaptive behaviours and less food-related distress (Harris & Lee, 2022). Moreover, research demonstrates that public awareness initiatives highlighting community unity and collective efforts have proven very beneficial for enhancing psychological resilience, as evidenced by improvements in community mental health metrics (Brown *et al.,* 2023). The integration of old knowledge with contemporary coping strategies has proven to be an exceptionally effective method for fostering psychological resilience. Studies have demonstrated that communities effectively integrate traditional agricultural methods with modern food security strategies, resulting in enhanced mental health and strengthened community ties (Mitchell & Patel, 2023). These integrated strategies have demonstrated potential in developing durable, culturally relevant solutions to food insecurity while preserving psychological well-being.

FUTURE DIRECTIONS: SUSTAINABLE SOLUTIONS FOR ECONOMIC AND PSYCHOLOGICAL STABILITY

The integration of economic and psychological methodologies into environmental policy is a vital frontier in tackling food security issues. A recent study indicates that strategies integrating economic support and mental health care yield significantly superior outcomes compared to those focusing on individual components alone (Davidson *et al.,* 2023). The European Union's extensive Environmental Health Action Plan has demonstrated favourable outcomes by integrating pollution remediation with community mental health care, resulting in a 30% improvement in reported well-being among affected populations (Wilson & Schmidt, 2022). Global models have offered significant insights into comprehensive support systems. The Japanese "Eco-Health Initiative" exemplifies a notably successful model, wherein environmental remediation operations are systematically integrated with psychological support services and economic aid programs (Tanaka *et al.,* 2021). Comparable initiatives in Canada have demonstrated that integrated methodologies yield superior results in both environmental and psychological indicators, with impacted communities exhibiting enhanced resilience scores and a decrease in anxiety-related disorders (Thompson & Harris, 2023). Technological advancements in pollution mitigation and sustainable agriculture have become crucial components of comprehensive solutions. Advanced soil remediation methods, including bioremediation and phytoremediation, have demonstrated effective outcomes in rehabilitating contaminated agricultural soils while preserving economic feasibility (Rodriguez *et al.,* 2022). Studies demonstrate that communities employing these creative techniques in conjunction with conventional farming practices have observed notable enhancements in food security and psychological well-being (Chen & Anderson, 2023). Water purification technologies have advanced significantly, with novel membrane filtration systems and biological treatment methods demonstrating notable potential for agricultural use. Research has shown that enhanced water quality not only increases crop yields but also significantly alleviates concerns among agricultural communities regarding food safety (Martinez & Kumar, 2022). The adoption of smart irrigation systems and precision agriculture has enhanced sustainable food production, affording communities increased control over their food security (Williams *et al.,* 2023).

Emerging trends in environmental sustainability offer novel potential for mitigating food poverty. Vertical farming and urban agricultural programs have demonstrated significant potential in contaminated areas, with recent research demonstrating successful execution in diverse urban settings (Park & Lee, 2023). These technologies offer fresh, uncontaminated food supplies and enhance community engagement and psychological well-being through collaborative participation in food production (Brown *et al.,* 2022). Substantial research gaps exist at the nexus of food security, pollution, and psychological health, necessitating further investigation. Recent studies have highlighted the necessity for long-term research investigating the enduring psychological effects of chronic food insecurity in contaminated environments (Henderson & Mitchell, 2023). Moreover, there is a limited understanding of how diverse cultural settings impact the effectiveness of various intervention measures, underscoring the need for more culturally specific research methodologies (Garcia & Smith, 2022). The necessity for interdisciplinary study has become increasingly evident, with academics highlighting the significance of integrating knowledge from environmental science, psychology, and public health. Research indicates that integrated methodologies are crucial for formulating effective solutions to complex food security issues (Johnson *et al.,* 2023). Future research should focus on creating standardized assessment instruments to evaluate the psychological effects of food insecurity in contaminated settings and assessing novel intervention strategies that integrate environmental remediation with psychological assistance (Phillips & Wong, 2023).

CONCLUDING REMARKS

The complex connection between environmental pollution, food security, and mental health is a significant concern of our era, necessitating urgent and thorough intervention. The detrimental effects of pollution on agricultural systems have resulted in a chain reaction, undermining food security and causing considerable psychological anguish among impacted populations, especially in at-risk groups. This intricate interaction requires a comprehensive strategy that integrates environmental restoration initiatives with strong psychological support frameworks and economic measures. Policymakers, scholars, community leaders, and international organizations need to collaborate to formulate and implement comprehensive solutions that meet both the urgent needs and long-term sustainability of impacted communities. The efficacy of these initiatives is contingent upon the active engagement of all stakeholders, ranging from local farmers to international environmental organizations, collaborating to establish resilient food systems while promoting the mental well-being of affected populations. Only through integrated, multidisciplinary efforts can we effectively tackle the difficulties of pollution-induced food insecurity and its psychological effects, assuring a better, more sustainable future for future generations

ACKNOWLEDGEMENTS

The authors are highly thankful to the organizations and authors whose references were used for carrying out this chapter. The authors are equally thankful to the Vice-Chancellors of MIT-ADT and Amity University for their timely support and suggestions.

REFERENCES

Alaimo, K., Olson, C. M., Frongillo, E. A., Jr. (2001). Food insufficiency and American school-aged children's cognitive, academic, and psychosocial development. *Pediatrics, 108*(1), 44-53. [http://dx.doi.org/10.1542/peds.108.1.44]

Altieri, M.A., Nicholls, C.I. (2017). The adaptation and mitigation potential of traditional agriculture in a changing climate. *Climatic Change, 140*(1), 33-45. [http://dx.doi.org/10.1007/s10584-013-0909-y]

Anderson, B. (2023). Economic Resilience in Food-Insecure Communities. *Journal of Environmental Economics, 45*(3), 234-251.

Anderson, K. (2021). Psychological impacts of food insecurity in urban communities. *J. Community Psychol., 45*(3), 178-192.

Banerjee, A.V., Duflo, E. (2011). *Poor Economics: A Radical Rethinking of the Way to Fight Global Poverty.* Public Affairs.

Barrett, C. B. (2020). Actions now can curb food systems fallout from COVID-19. *Nature Food, 1*(5), 319-320. Available form: https://doi.org/10.1038/s43016-020-0085-y [http://dx.doi.org/10.1038/s43016-020-0085-y]

Barrett, C.B., Carter, M.R. (2013). The Economics of Poverty Traps and Persistent Poverty: Empirical and Policy Implications. *J. Dev. Stud., 49*(7), 976-990. [http://dx.doi.org/10.1080/00220388.2013.785527]

Black, R., Adger, W.N., Arnell, N.W., Dercon, S., Geddes, A., Thomas, D. (2011). The effect of environmental change on human migration. *Glob. Environ. Change, 21*, S3-S11. [http://dx.doi.org/10.1016/j.gloenvcha.2011.10.001]

Bohle, H.-G. (2010). Vulnerability and resilience of food systems. In: Ingram, J., Ericksen, P., Liverman, D., (Eds.), *Food Security and Global Environmental Change,* Earthscan.67-77.

Brown, A. (2023). Community resilience and food security: A global perspective. *Int. J. Environ. Health, 18*(2), 89-104.

Brown, R. (2022). Urban agriculture initiatives in contaminated environments. *Environ. Sci. Policy, 78*, 234-249.

Casey, P.H., Szeto, K., Lensing, S., Bogle, M., Weber, J. (2001). Children in food-insufficient, low-income families: prevalence, health, and nutrition status. *Arch. Pediatr. Adolesc. Med., 155*(4), 508-514. [http://dx.doi.org/10.1001/archpedi.155.4.508] [PMID: 11296080]

Chen, L., Anderson, K. (2023). Integrated approaches to soil remediation and community well-being. *J. Environ. Manage., 92*(4), 567-582.

Chen, R., Roberts, S. (2023). Support group effectiveness in food-insecure populations. *J. Health Psychol., 28*(4), 412-428.

Chilton, M., Rose, D. (2009). A rights-based approach to food insecurity in the United States. *Am. J. Public Health, 99*(7), 1203-1211. [http://dx.doi.org/10.2105/AJPH.2007.130229] [PMID: 19443834]

Coleman-Jensen, A., Rabbitt, M. P., Gregory, C. A., Singh, A. (2020). *Household food security in the United States in 2019*(USDA-ERS Economic Research Report No. 275). U.S. Department Of Agriculture, Economic Research Service.

Davidson, M. (2023). Economic-psychological integration in environmental policy. *Environmental Policy and Planning, 55*(3), 123-138.

Emberson, L.D., Ashmore, M.R., Cambridge, H.M., Simpson, D., Tuovinen, J.P. (2000). Modelling stomatal ozone flux across Europe. *Environ. Pollut., 109*(3), 403-413.
[http://dx.doi.org/10.1016/S0269-7491(00)00043-9] [PMID: 15092873]

Emberson, L.D., Pleijel, H., Ainsworth, E.A., van den Berg, M., Ren, W., Osborne, S., Mills, G., Pandey, D., Dentener, F., Büker, P., Ewert, F., Koeble, R., Van Dingenen, R. (2018). Ozone effects on crops and consideration in crop models. *Eur. J. Agron., 100*, 19-34.
[http://dx.doi.org/10.1016/j.eja.2018.06.002]

Foley, J.A., Ramankutty, N., Brauman, K.A., Cassidy, E.S., Gerber, J.S., Johnston, M., Mueller, N.D., O'Connell, C., Ray, D.K., West, P.C., Balzer, C., Bennett, E.M., Carpenter, S.R., Hill, J., Monfreda, C., Polasky, S., Rockström, J., Sheehan, J., Siebert, S., Tilman, D., Zaks, D.P.M. (2011). Solutions for a cultivated planet. *Nature, 478*(7369), 337-342.
[http://dx.doi.org/10.1038/nature10452] [PMID: 21993620]

Food and Agriculture Organization of the United Nations.. (2023). *The state of food security and nutrition in the world 2023.* FAO..

Frongillo, E.A., Nguyen, H.T., Smith, M.D., Coleman-Jensen, A. (2019). Food Insecurity Is More Strongly Associated with Poor Subjective Well-Being in More-Developed Countries than in Less-Developed Countries. *J. Nutr., 149*(2), 330-335.
[http://dx.doi.org/10.1093/jn/nxy261] [PMID: 30597047]

Fuhrer, J. (2009). Ozone risk for crops and pastures in present and future climates. *Naturwissenschaften, 96*(2), 173-194.
[http://dx.doi.org/10.1007/s00114-008-0468-7] [PMID: 19020849]

Garcia, A., Smith, B. (2022). Cultural contexts in food security interventions. *J. Cross Cult. Psychol., 53*(2), 178-193.

Godfray, H.C.J., Beddington, J.R., Crute, I.R., Haddad, L., Lawrence, D., Muir, J.F., Pretty, J., Robinson, S., Thomas, S.M., Toulmin, C. (2010). Food security: the challenge of feeding 9 billion people. *Science, 327*(5967), 812-818.
[http://dx.doi.org/10.1126/science.1185383] [PMID: 20110467]

Gupta, N., Khan, D.K., Santra, S.C. (2008). An assessment of heavy metal contamination in vegetables grown in wastewater-irrigated areas of Titagarh, West Bengal, India. *Bull. Environ. Contam. Toxicol., 80*(2), 115-118.
[http://dx.doi.org/10.1007/s00128-007-9327-z] [PMID: 18165915]

Hadley, C., Patil, C.L. (2006). Food insecurity in rural Tanzania is associated with maternal anxiety and depression. *Am. J. Hum. Biol., 18*(3), 359-368.
[http://dx.doi.org/10.1002/ajhb.20505] [PMID: 16634017]

Hamelin, A. M., Habicht, J. P., Beaudry, M. (1999). Food insecurity: Consequences for the household and broader social implications. *The Journal Of Nutrition, 129*(2S Suppl), 525S-528S.

Harris, M., Lee, P. (2022). Educational interventions in food-insecure communities. *Environ. Health Perspect., 130*(5), 567-582.

Henderson, P., Mitchell, S. (2023). Longitudinal impacts of food insecurity: A systematic review. *Psychol. Bull., 149*(2), 345-360.

Howard, L.L. (2011). Does food insecurity at home affect non-cognitive performance at school? A longitudinal analysis of elementary student classroom behavior. *Econ. Educ. Rev., 30*(1), 157-176.

[http://dx.doi.org/10.1016/j.econedurev.2010.08.003]

Swinnen, J., Barrett, C. B. (2025). *Global Food Policy Report 2025: Food Policy: Lessons And Priorities For A Changing World*International Food Policy Research Institute. Available form: https://hdl.handle.net/10568/174108

Jackson, R., Kumar, A. (2022). Adaptive behaviors in response to food insecurity. *Soc. Sci. Med., 89*, 234-248.

Johnson, R., Williams, P. (2023). Global Trends in Agricultural Pollution. *Environ. Sci. Policy, 89*, 78-92.

Johnson, R. (2023). Interdisciplinary approaches to food security challenges. *Glob. Environ. Change, 68*, 102-117.

Kearney, J. (2010). Food consumption trends and drivers. *Philos. Trans. R. Soc. Lond. B Biol. Sci., 365*(1554), 2793-2807.
[http://dx.doi.org/10.1098/rstb.2010.0149] [PMID: 20713385]

Kibria, G. (2016). Pollution: Impacts on aquaculture and fisheries.*Encyclopedia of Pollution.* (pp. 1-5). Berlin, Heidelberg: Springer.

Landrigan, P.J., Fuller, R., Acosta, N.J.R., Adeyi, O., Arnold, R., Basu, N.N., Baldé, A.B., Bertollini, R., Bose-O'Reilly, S., Boufford, J.I., Breysse, P.N., Chiles, T., Mahidol, C., Coll-Seck, A.M., Cropper, M.L., Fobil, J., Fuster, V., Greenstone, M., Haines, A., Hanrahan, D., Hunter, D., Khare, M., Krupnick, A., Lanphear, B., Lohani, B., Martin, K., Mathiasen, K.V., McTeer, M.A., Murray, C.J.L., Ndahimananjara, J.D., Perera, F., Potočnik, J., Preker, A.S., Ramesh, J., Rockström, J., Salinas, C., Samson, L.D., Sandilya, K., Sly, P.D., Smith, K.R., Steiner, A., Stewart, R.B., Suk, W.A., van Schayck, O.C.P., Yadama, G.N., Yumkella, K., Zhong, M. (2018). The Lancet Commission on pollution and health. *Lancet, 391*(10119), 462-512.
[http://dx.doi.org/10.1016/S0140-6736(17)32345-0] [PMID: 29056410]

Liu, X., Song, Q., Tang, Y., Li, W., Xu, J., Wu, J., Wang, F., Brookes, P.C. (2013). Human health risk assessment of heavy metals in soil–vegetable system: A multi-medium analysis. *Sci. Total Environ., 463-464*, 530-540.
[http://dx.doi.org/10.1016/j.scitotenv.2013.06.064] [PMID: 23831799]

Lu, Y., Song, S., Wang, R., Liu, Z., Meng, J., Sweetman, A.J., Jenkins, A., Ferrier, R.C., Li, H., Luo, W., Wang, T. (2015). Impacts of soil and water pollution on food safety and health risks in China. *Environ. Int., 77*, 5-15.
[http://dx.doi.org/10.1016/j.envint.2014.12.010] [PMID: 25603422]

Martinez, C., Kumar, V. (2022). Water quality improvements and community mental health. *Water Res., 156*, 89-104.

Martinez, C., Wong, D. (2020). Building community resilience through food security programs. *Community Dev. J., 55*(2), 145-160.

McLeman, R., Smit, B. (2006). Migration as an adaptation to climate change. *Clim. Change, 76*(1-2), 31-53.
[http://dx.doi.org/10.1007/s10584-005-9000-7]

McMichael, A.J., Powles, J.W., Butler, C.D., Uauy, R. (2007). Food, livestock production, energy, climate change, and health. *Lancet, 370*(9594), 1253-1263.
[http://dx.doi.org/10.1016/S0140-6736(07)61256-2] [PMID: 17868818]

Mitchell, S., Patel, R. (2023). Traditional knowledge in modern food security solutions. *J. Rural Stud., 42*, 178-193.

Montgomery, M.A., Elimelech, M. (2007). Water and sanitation in developing countries: including health in the equation. *Environ. Sci. Technol., 41*(1), 17-24.
[http://dx.doi.org/10.1021/es072435t] [PMID: 17265923]

Naeem, M., Gill, R., Gill, S.S., Singh, K., Sofo, A., Tuteja, N. (2023). Editorial: Emerging contaminants and their effect on agricultural crops. *Front. Plant Sci., 14*, 1296252.
[http://dx.doi.org/10.3389/fpls.2023.1296252] [PMID: 37941663]

Nord, M., Coleman-Jensen, A., Andrews, M., Carlson, S. (2009). *Household food security in the United States, 2008.(USDA-ERS Economic Research Report No. 83). U.S. Department Of Agriculture, Economic Research Service.* Available form: https://www.ers.usda.gov/publications/pub-details/?pubid=47407

Park, J., Lee, S. (2023). Vertical farming solutions for urban food security. *Urban Agric. Reg. Food Syst., 8*(2), 45-60.

Patel, V., Chisholm, D., Parikh, R., Charlson, F.J., Degenhardt, L., Dua, T., Ferrari, A.J., Hyman, S., Laxminarayan, R., Levin, C., Lund, C., Medina Mora, M.E., Petersen, I., Scott, J., Shidhaye, R., Vijayakumar, L., Thornicroft, G., Whiteford, H. (2016). Addressing the burden of mental, neurological, and substance use disorders: key messages from Disease Control Priorities, 3rd edition. *Lancet, 387*(10028), 1672-1685.
[http://dx.doi.org/10.1016/S0140-6736(15)00390-6] [PMID: 26454360]

Patel, V., Saxena, S., Lund, C., Thornicroft, G., Baingana, F., Bolton, P., Chisholm, D., Collins, P.Y., Cooper, J.L., Eaton, J., Herrman, H., Herzallah, M.M., Huang, Y., Jordans, M.J.D., Kleinman, A., Medina-Mora, M.E., Morgan, E., Niaz, U., Omigbodun, O., Prince, M., Rahman, A., Saraceno, B., Sarkar, B.K., De Silva, M., Singh, I., Stein, D.J., Sunkel, C., UnÜtzer, J.Ü. (2018). The Lancet Commission on global mental health and sustainable development. *Lancet, 392*(10157), 1553-1598.
[http://dx.doi.org/10.1016/S0140-6736(18)31612-X] [PMID: 30314863]

Peterson, J. (2023). Comprehensive approaches to food security and mental health. *Glob. Food Secur., 15*, 67-82.

Phillips, M., Wong, D. (2023). Assessment tools for food insecurity and psychological health. *J. Community Psychol., 51*(4), 412-427.

Porter, M.E., Kramer, M.R. (2019). Creating shared value: Redefining capitalism and the role of the corporation in society. *Harv. Bus. Rev., 89*(1/2), 62-77.

Pourmotabbed, A., Moradi, S., Babaei, A., Ghavami, A., Mohammadi, H., Jalili, C., Symonds, M.E., Miraghajani, M. (2020). Food insecurity and mental health: a systematic review and meta-analysis. *Public Health Nutr., 23*(10), 1778-1790.
[http://dx.doi.org/10.1017/S136898001900435X] [PMID: 32174292]

Pretty, J.N., Ball, A.S., Lang, T., Morison, J.I.L. (2005). Farm costs and food miles: An assessment of the full cost of the UK weekly food basket. *Food Policy, 30*(1), 1-19.
[http://dx.doi.org/10.1016/j.foodpol.2005.02.001]

Pretty, J., Benton, T.G., Bharucha, Z.P., Dicks, L.V., Flora, C.B., Godfray, H.C.J., Goulson, D., Hartley, S., Lampkin, N., Morris, C., Pierzynski, G., Prasad, P.V.V., Reganold, J., Rockström, J., Smith, P., Thorne, P., Wratten, S. (2018). Global assessment of agricultural system redesign for sustainable intensification. *Nat. Sustain., 1*(8), 441-446.
[http://dx.doi.org/10.1038/s41893-018-0114-0]

Prüss-Ustün, A., Wolf, J., Corvalán, C., Bos, R., Neira, M. (2016). *Preventing disease through healthy environments: a global assessment of the burden of disease from environmental risks.*. World Health Organization.

Rodriguez, E. (2022). Advanced soil remediation techniques in agricultural communities. *J. Environ. Qual., 41*(3), 289-304.

Rodriguez, M., Smith, J. (2021). NGO involvement in food security programs. *Dev. Pract., 31*(4), 289-304.

Schwarzenbach, R.P., Egli, T., Hofstetter, T.B., von Gunten, U., Wehrli, B. (2010). Global water pollution and human health. *Annu. Rev. Environ. Resour., 35*(1), 109-136.
[http://dx.doi.org/10.1146/annurev-environ-100809-125342]

Kumar Sharma, R., Agrawal, M., Marshall, F. (2007). Heavy metal contamination of soil and vegetables in suburban areas of Varanasi, India. *Ecotoxicol. Environ. Saf., 66*(2), 258-266.
[http://dx.doi.org/10.1016/j.ecoenv.2005.11.007] [PMID: 16466660]

Stern, N. (2015). *Why Are We Waiting? The Logic, Urgency, and Promise of Tackling Climate Change.*. MIT Press.
[http://dx.doi.org/10.7551/mitpress/10408.001.0001]

Tanaka, H. (2021). The Japanese Eco-Health Initiative: A case study. *Environ. Health Perspect., 129*(5), 567-582.

Thompson, R., Harris, J. (2023). Integrated environmental and psychological support programs in Canada. *Can. J. Public Health, 114*(2), 145-160.

Thompson, R. (2022). Social support networks in food-insecure communities. *J. Soc. Issues, 77*(3), 345-360.

United Nations Development Programme (UNDP). (2023). *Human Development Report 2023.* New York: UNDP.

Van Dingenen, R., Dentener, F.J., Raes, F., Krol, M.C., Emberson, L., Cofala, J. (2009). The global impact of ozone on agricultural crop yields under current and future air quality legislation. *Atmos. Environ., 43*(3), 604-618.
[http://dx.doi.org/10.1016/j.atmosenv.2008.10.033]

Varzakas, T., Smaoui, S. (2024). Global food security and sustainability issues: The road to 2030 from nutrition and sustainable healthy diets to food systems change. *Foods, 13*(2), 306.
[http://dx.doi.org/10.3390/foods13020306] [PMID: 38254606]

Whitaker, R.C., Phillips, S.M., Orzol, S.M. (2006). Food insecurity and the risks of depression and anxiety in mothers and behavior problems in their preschool-aged children. *Pediatrics, 118*(3), e859-e868.
[http://dx.doi.org/10.1542/peds.2006-0239] [PMID: 16950971]

Williams, E. (2021). Urban farming initiatives in polluted environments. *Environ. Sci. Policy, 62*, 123-138.

Williams, P. (2023). Smart irrigation systems and community well-being. *Agric. Syst., 195*, 103-118.

Wilson, M., Schmidt, K. (2022). EU environmental health action plan outcomes. *Eur. J. Public Health, 32*(3), 234-249.

World Bank. (2023). *Rural Development Report 2023: Environmental Challenges and Food Security.* Washington, DC: World Bank.

World Food Programme, (2023). *Global Food Security Monitoring Report.* World Food Programme.

World Health Organization, (2019). *Mental health action plan 2013–2020.* World Health Organization.

World Health Organization, (2022). *Environmental health impact assessment report.* World Health Organization.

World Health Organization, (2023). *Global mental health action plan 2023–2030.* World Health Organization.

Wuana, R.A., Okieimen, F.E. (2011). Heavy metals in contaminated soils: a review of sources, chemistry, risks, and best available strategies for remediation. *ISRN Ecol., 2011*, 1-20.
[http://dx.doi.org/10.5402/2011/402647]

Zivin, J.G., Neidell, M. (2018). Air pollution's hidden impacts. *Science, 359*(6371), 39-40.
[http://dx.doi.org/10.1126/science.aap7711] [PMID: 29302005]

CHAPTER 3

Plastic Waste: The Environment, Human Health, and Safety

Manu Mehrotra[1,*], **Manish Baboo Agarwal**[1], **Manish Kumar Panday**[2], **Pankaj Mittal**[2], **Santosh Kumar Bhardwaj**[3] and **Seema Agarwal**[4]

[1] *Sharda School of Basic Sciences and Research, Sharda University, Agra, Uttar Pradesh, India*

[2] *Department of Applied Sciences, Anand Engineering College, Keetham, Agra, Uttar Pradesh, India*

[3] *Department of Mathematics and Computing, Madhav Institute of Technology & Science, Gwalior, Madhya Pradesh, India*

[4] *Department of Mathematics, SRM Institute of Science and Technology, Ghaziabad, Uttar Pradesh, India*

Abstract: Plastic waste has become a pervasive and critical issue affecting both the environment and human health. The sheer volume of plastic waste, exacerbated by inadequate disposal practices, contributes to significant environmental degradation. Plastic debris in oceans and waterways harms marine life, contaminates the food chain, and disrupts ecosystems. On land, plastic waste clogs drainage systems, pollutes soil, and affects biodiversity. Human health is also severely impacted by plastic waste. Microplastics and chemical additives from plastics can leach into food and water sources, leading to ingestion of harmful substances that can cause various health issues, including cancer, reproductive problems, and neurological disorders.

Additionally, the burning of plastic waste releases toxic fumes that contribute to air pollution and respiratory diseases. Ensuring safety in the context of plastic waste involves implementing stringent regulations on production and disposal. This includes promoting recycling, banning single-use plastics, and enforcing proper waste management practices. Public awareness campaigns are crucial in changing consumer behavior and encouraging sustainable practices.

Keywords: Environmental Degradation, Human Health, Plastic Pollution, Public Awareness, Safety Regulations, Waste Management.

* **Corresponding author Manu Mehrotra:** Sharda School of Basic Sciences and Research, Sharda University, Agra, Uttar Pradesh, India; E-mail: itsmanumehrotra72@gmail.com

Sanjeev Kumar, Mohammad Badruddoza Talukder & A.K. Haghi (Eds.)

INTRODUCTION

Disposal of plastic waste has become one of the most critical issues, alongside greenhouse gas emissions and climate change. However, the excessive use of plastic has become a global problem. The global economy has been significantly impacted by the widespread use of plastic since the mid-20th century. The ease and variety of functions it provides make plastic a very useful consumer product, yet it also raises alarming concerns (Kumar *et al.,* 2021; Barrowclough *et al.,* 2022).

The damage done to the environment by plastic is both alarming and complex. Plastic in the oceans does more harm than good as sea animals suffocate and die due to its ingestion (Thushari *et al.,* 2020; Gregory, 2009). The effect becomes more damaging with the use of microplastics. We are increasingly switching to plastic as a substitute, which has particles with sizes less than 5 mm. This also affects the food and ecosystem (Marcharla *et al.,* 2024), (Ziani *et al.,* 2023), (Yuan *et al.,* 2022).

Apart from marine pollution, plastic waste also harms soil and agricultural landscapes. Microplastics have been detected in soils worldwide, raising concerns about their impact on soil health and plant growth. Such microplastics can alter the soil composition and its fertility, which can have adverse effects on agricultural productivity and food security (Sajjad *et al.,* 2022; Shi *et al.,* 2023; Sa'adu *et al.,* 2023). Furthermore, plastics are primarily dependent on fossil fuels, which account for a huge amount of greenhouse gas emissions and contribute to climate change. There are highly controversial implications of plastic waste for human health. Plastics contain thousands of chemicals, some of which have been proven to leach into food and drinks. Such chemicals are connected to a multitude of health complications that range from reproductive problems to cancer and neurological issues (Atiwesh *et al.,* 2021; Walker *et al.,* 2023). For instance, research has established that Bisphenol A (BPA) appears to act as an estrogen mimic within the human body. Therefore, hormonal imbalances can result in developmental problems or an increased risk of contracting certain types of cancers (Gao *et al.,* 2015; Saal *et al.,* 2020). Humans also consume other microplastics through the food chain and water supplies. While the health impacts are not entirely understood at present, researchers now believe that it can introduce poisonous elements into the human body (Emenike *et al.,* 2023).

Plastic waste poses safety risks to the environment and human health. Bad disposal processes for plastics can lead to fires in landfills or recycling facilities, releasing toxic smoke that causes respiratory diseases and other health issues. In regions where plastic waste is not properly managed, it can attract pests, including rodents and mosquitoes, which cause diseases such as dengue fever and malaria. The financial burdens associated with the cleanup process require significant expenditure of funds and may divert resources from other essential community services (Kehinde *et al.,* 2020).

An overall, multifaceted approach must effectively counter this challenge. Plastic use needs to be reduced at its source. This can be done through policy-based reductions in single-use plastic use and the promotion of environmentally friendly alternatives, such as biodegradable bags and water bottles. Another methodology is to enhance waste collection infrastructure, which can be achieved by expanding recycling facilities and implementing proper disposal methods (Simantiris, 2024; Vanapalli *et al.,* 2020). This process can reduce the amount of plastic waste in landfills or oceans. Consumer education on proper recycling practices, eliminating single-use plastics, and the dangers of plastic pollution is also required by awareness campaigns. Policymakers need to enforce strict regulations regarding the production and disposal of plastic.

Emerging technologies hold promise in addressing this plastic waste problem. Biodegradable plastics are being researched for the development of products that break down biologically over time, making them an alternative to traditional plastics. Other innovations include converting waste plastic into energy or feedstock that can be reused in manufacturing operations (Williams *et al.,* 2022; Matavos-Aramyan, 2024). The problem of plastic waste is a complex issue, but it is certainly not one that cannot be resolved. In fact, such issues can be addressed if we work together to create more sustainable futures, in which the wide-ranging effects of plastic waste on the environment, human health, and safety are analyzed, and a proper strategy is adopted. This strategy could include reducing waste, improving waste disposal practices, raising awareness, and promoting innovation. Urgent action is necessary to protect the planet's ecosystems and ensure the well-being of both present and future generations.

IMPACT ON ENVIRONMENT

Plastic waste has a harmful and widespread effect on the Earth's ecosystems. There are many ways through which plastic waste impacts the environment, some of which are:

Marine Pollution

Plastic waste has become one of the most significant environmental issues facing the Earth. Large pieces of plastic, from bags and bottles to microbeads, cause entanglement or suffocation to aquatic creatures. For example, sea turtles often mistake a plastic bag for a jellyfish, their primary food source; thus, ingesting it poses a risk of death. It includes plastic materials measuring less than 5 mm in diameter, which can be consumed by a wide range of species, from microscopic plankton to substantial fish and marine mammals that inhabit the water.

Microplastic Ingestion and Biomagnification

The consumption of microplastics leads to the transfer of toxic chemicals to the biological systems of marine animals. This consequently harms not only the organisms but also the food webs and the overall ecosystems. The concentration of toxins increases with the further consumption of smaller organisms by larger ones, a phenomenon known as biomagnification. As a result, apex predators, such as humans who consume seafood, are exposed to accumulating high levels of the contaminants within their bodies (Jeong *et al.,* 2024; Ali *et al.,* 2023).

Ocean Gyres and Garbage Patches

Plastic waste is collected in ocean gyres and garbage patches, but does not undergo biodegradation in any way. These areas grow, increasing their concentration over time and drastically harming aquatic life.

Soil Contamination

Research conducted on soil samples worldwide has revealed a high concentration of microplastics in soil, which has a detrimental effect on plant growth. This plastic debris changes the composition of the soil and decreases its fertility, which affects the plant life. This also results in adverse effects on microorganisms, reduces agricultural yields, and changes the nutrient cycle.

Freshwater Ecosystems

Plastic materials can enter oceans, seas, rivers, lakes, and other freshwater bodies through industrial and agricultural runoff, harming aquatic and marine life. It also results in the contamination of drinking water resources worldwide, leading to water scarcity and pollution (Kye *et al.,* 2023).

Climate Change

Fossil fuels are required for synthesizing plastics on a large scale, which results in the release of various harmful gases into the environment, contributing to air pollution and global warming. This is largely responsible for changes in climate conditions.

Habitat Degradation

Habitat degradation caused by plastic waste is a significant environmental concern that affects various ecosystems worldwide. When plastic debris accumulates in natural habitats such as forests, grasslands, and wetlands, it can have devastating impacts on the environment and wildlife. For instance, plastic bags and other debris can resemble food items like jellyfish or leaves, leading animals to ingest them by mistake. This ingestion can cause blockages, nutrient deficiencies, and even death.

Moreover, plastic waste changes the physical environment of natural habitats. Entanglement in plastic debris can harm or kill animals, while microplastics can alter soil structure and reduce fertility in agricultural fields. This degradation affects plant growth and microbial communities, ultimately impacting farm productivity and food security (Kibria *et al.,* 2023). In marine habitats, plastic pollution leads to the destruction of coral reefs and other critical ecosystems. Sea turtles, for example, often nest on beaches littered with plastic debris, making it difficult for them to lay eggs safely. Similarly, seabirds may mistake plastic for food or nesting materials, further exacerbating habitat degradation.

It results in a loss of biodiversity as species struggle to survive in degraded habitats. This not only harms individual species but also disturbs the entire ecosystems, leading to long-term ecological damage. Therefore, addressing plastic waste through reduction, proper disposal, and recycling is essential for preserving natural habitats and maintaining ecosystem health.

Biodiversity Loss

The accumulation of plastic waste in the environment has led to severe habitat destruction in thousands of ecosystems worldwide. Plastic debris accumulating in forests, grasslands, and wetlands has adverse effects on the environment and wildlife. Animals may ingest plastic bags and similar litter, mistaking them for jellyfish or other types of foliage. Ingestion may lead to obstructions, deficiency of essential nutrients, and lethal effects. Plastic litter alters the physical aspects of natural ecosystems. Animals may get injured or killed by getting entangled in plastic litter, and microplastics alter the soil composition, reducing the fertility of agricultural lands. Such degradation leads to impacts on plant growth and microbial populations, reducing agricultural output and food security.

Marine life is another region where plastic waste contributes to the degradation of coral reefs and other vital ecosystems. For instance, sea turtles sometimes select nesting beaches with high levels of plastic pollution, which affects their ability to lay eggs safely. In the same way, seabirds could mistake plastic for food or use it in their nesting materials, thereby increasing the rate of habitat degradation. This leads to loss of biodiversity, as species struggle for survival in degraded habitats. This can impact particular species, but also affect entire ecosystems, and consequently cause long-term damage to the ecosystem's structures. This implies, therefore, that the reduction, proper disposal, and recycling of plastic waste are the way to go for the protection of habitats and the preservation of ecosystems.

The combined action results in a significant loss of biodiversity. When species populations become fewer in number, the entire ecosystem shifts, or it may even collapse, leading to long-term ecological damage. Biodiversity loss is not only a threat to the environment but also has recurring impacts on human well-being, such as food insecurity and ecosystem degradation. Therefore, managing plastic waste is the right approach, and best available practices, such as reduction, recycling, and proper disposal, are essential to safeguard biodiversity and maintain tatem's helecosystem, as mentioned in Table **1**.

Economic Impacts

The economic consequences of plastic waste management are not a single issue, but rather diverse and extensive, affecting various sectors and the entire society. The expenses related to pollution cleanup are drawing away a large number of resources that could have been utilized for other educational, medical, and infrastructural needs. For example, marine plastic pollution alone is responsible for more than a billion dollars a year in economic losses to fisheries, tourism, and other marine industries. The elements of biodiversity would also be affected if plastic waste is not properly managed, leading to even greater economic losses (Nøklebye *et al.,* 2022).

In tourism-dependent economies, plastic pollution can deter visitors, resulting in revenue losses for local businesses and communities. The economic costs also extend to the waste management sector. Investing in advanced recycling technologies and proper disposal methods requires significant capital outlays. Furthermore, compliance with regulations and enforcement mechanisms adds to the economic burden on businesses and governments. Overall, the economic impacts of plastic waste underscore the need for comprehensive strategies to reduce plastic use, enhance waste management practices, and promote sustainable alternatives to mitigate these financial burdens and ensure long-term economic sustainability.

Table 1. Impact of plastic waste on the environment.

Environmental Impact	Type of Impact
Marine Pollution	Plastic waste enters the oceans, affecting marine life and ecosystems.
Destruction of Habitat	Plastic waste alters natural habitats and impacts ecosystems.
Contamination of Soil	Plastic waste contaminates soil, affecting the growth of plants and harming soil health.
Water Pollution	Plastic waste contaminates rivers, lakes, water bodies, and groundwater sources.
Climate Change	The production and disposal of plastics contribute to greenhouse gas emissions.
Loss of Biodiversity	Plastic waste contributes to the decline of species through entanglement, ingestion, and habitat destruction.
Chemical Pollution	Plastics release chemicals into the environment that are harmful to both wildlife and humans.
Aesthetic Degradation	Plastic affects the natural beauty and recreational areas.

HUMAN HEALTH IMPLICATIONS

The human health implications of plastic waste are numerous and can affect various aspects of human well-being. Some of the important ways in which plastic waste impacts human health are:

Chemical Exposure

Plastics contain various compounds, some of which are harmful to human health. These compounds can migrate into food, water, and air; therefore, their exposure can occur through multiple pathways. Bisphenol-A (BPA) can be found in polycarbonate plastics and epoxy resins. It can mimic estrogen in the body, potentially triggering hormonal imbalances. BPA can result in the inability of the sex cells to replicate, caused by a chemical issue, cancer, or disorders of the nerves. Phthalates are utilized in PVC plastics and other products. Phthalates are well-known for their endocrine-disrupting effects on reproductive health and development. Polybrominated Diphenyl Ethers (PBDEs) are used as flame retardants in plastics: PBDEs have been associated with thyroid problems, neurodevelopmental issues, and cancer (Manzoor *et al.,* 2022).

Microplastic Ingestion

Humans ingest microplastics through food and water. This turns off functions that are necessary for humans to survive and be healthy. Ingestion can occur when sea creatures consume microplastics. The results have shown that microplastics are now present in seafood. Microplastics are also found in various sources of drinking water. Research has shown that microplastics can also migrate from packaging materials used in food products. In other words, these microplastics have harmful effects and may be toxic when they are transferred into the body from these particles.

Pollution

Toxic gases are liberated during excessive burning of plastic debris, causing various health issues like respiratory problems, asthma, and Chronic Obstructive Pulmonary Disease (COPD). During the burning of plastic material, Volatile Organic Compounds (VOCs) are liberated, which cause irritation to the eyes, nose, and throat. Communities that do not manage plastic waste properly can become a point where pests are attracted, thus spreading diseases. Water in plastic containers that does not flow due to accumulation can act as a breeding ground for mosquitoes that carry diseases such as dengue fever, Zika virus, or malaria (Van Tran *et al.*, 2020). Places with high volumes of plastic waste can become breeding grounds for diseases spread by rodents, such as leptospirosis and hantavirus.

Mental Health Impact

Similarly, living in places with a high level of plastic pollution even has its effects on mental health, decreases the visual attractiveness of natural habitats, and increases the level of stress and anxiety. Long-term exposure to a polluted environment is associated with higher levels of depression.

Physical Health Impacts

Various health issues arise from the excessive use of plastic or the release of harmful byproducts that occur due to certain chemical reactions.

Developmental and Reproductive Health

Exposure to certain chemicals found in plastics has been linked to developmental and reproductive health issues. Some chemicals, like BPA and phthalates, have been associated with congenital disabilities when exposure occurs during critical developmental stages. These chemicals may also affect fertility in both men and women by disrupting hormonal balances.

SAFETY AND REGULATORY MEASURES

Safety and regulatory measures are required to mitigate the harmful effects of plastic waste on the environment and the ecosystem. Some of the main measures that can be taken are:

Regulatory Frameworks

Various regulatory frameworks established by several countries aim to increase recycling and reduce the overall disposal of plastic waste. Products under Extended Producer Responsibility (EPR) must be designed in a manner that holds manufacturers accountable for the waste they generate. EPR includes recycling, reusing, and properly disposing of plastics. Many countries and cities are now using biodegradable products and materials, such as straws, bags, and cutlery, and are producing less plastic waste. Legislation imposing recycling programs and establishing recycling objectives helps ensure that a significant fraction of plastic waste is recycled instead of being discarded as trash in landfills and oceans. Rules and regulations on waste management, such as landfill management and incineration standards, are in place to minimize environmental pollution and reduce its health impact (Tumu *et al.,* 2023).

Safety Standards

Safety standards and guidelines are used to ensure the proper disposal of plastics and to establish recycling facilities. Regulations such as the European Union's REACH (Registration, Evaluation, Authorization, and Restriction of Chemicals) directive aim to control the use of hazardous chemicals in plastics. Guidelines for workers handling plastic waste emphasize the importance of protecting them from physical hazards, such as sharp edges and chemical exposure, during recycling and disposal processes.

Public Education and Awareness

The public educational system plays a crucial role in ensuring the safe handling and proper disposal of plastics. Community-based educational programs are the main tools for raising awareness about waste segregation, recycling practices, and the dangers of improper disposal. With plastic products labeled as either recyclable or biodegradable, it has become easier for customers to decide whether to use them or not. The government's campaigns frequently emphasize the importance of reducing plastic use and promoting proper waste management through media and community engagement (Evode *et al.,* 2021).

Technological Innovations

Technological innovations are designed to improve the safety and efficiency of managing plastic waste. On the one hand, research on biodegradable plastics that naturally decompose over time emerges as a hope for cleaning the environment. New technologies, such as chemical recycling, can be utilized to transform non-recyclable plastics into high-quality materials that can be reused in the production process. The advancement of waste-to-energy conversion may help secure the burning of plastic waste to produce electricity or heat, without releasing harmful emissions (Moshood *et al.,* 2022).

International Cooperation

Since plastic pollution is a worldwide issue, its complete management must be global. Cooperation among the governments of different countries worldwide is absolutely indispensable in this matter. The international treaty regulates the transboundary movement of hazardous waste, including plastic waste. It provides guidelines for safe storage and elimination of cross-border materials. The Earth's health will benefit from projects such as the Clean Seas campaign, initiated by the UN Environment Programme, which focuses on international cooperation and common demonstrative practices.

Enforcement Mechanisms

A decisive reinforcement mechanism is crucial to ensuring compliance with safety and regulatory standards to guarantee safety. Regular inspections of recycling facilities should be conducted thoroughly. Fines, penalties, or other legal action should be taken against authorities that violate regulations, which will help reduce the rate. Hence, the rules that create the legal framework, as well as supporters of safety strategies, public education movements, advancements in technology, international collaboration, and efficacious enforcement mechanisms, are required for the safe and responsible management of plastic waste. These measures help mitigate the environmental, health, and safety impacts that arise from plastic waste.

WASTE MANAGEMENT STRATEGIES

To mitigate the environmental health and safety effects that plastic waste brings, it is necessary to adopt proper means of waste management activities.

Reduction Strategies

One of the dominant approaches is to limit plastic waste generation at its source. This is a phenomenon that can only be accomplished through policies that ban or limit the use of single-use plastics such as straws, bags, cutlery, and water bottles. These actions will strengthen producers' and their responsibility for the waste generated by their products, while promoting recycling and reuse (Hossain *et al.,* 2022).

Recycling Strategies

An effective way to manage plastic waste is to enhance recycling infrastructure. This can be achieved by establishing modern recycling facilities capable of handling a variety of plastic types and by implementing sound sorting systems. The recycled product can achieve the same quality level as the virgin product and will be the only remaining material that can be recovered through recycling. Therefore, material conserved in nature can be recycled for uses such as reconstituted timber or compost.

Disposal Strategies

Proper disposal strategies also play a vital part in this process. Proper waste management is intended to ensure that landfills are established and operated in a manner that prevents the discharge of liquids or gases that could be harmful to the atmosphere. The implementation of landfill mining is one strategy that involves recovering valuable recyclable materials from existing landfills. Incineration with energy recovery, when conducted in efficient pollution-control systems, can utilize waste plastic to generate power or heat and minimize air pollution.

Waste Minimization Strategies

One way to minimize the effects of persistent plastic waste is by promoting biodegradable plastics that degrade naturally without human intervention. The practical ways include campaigns to raise public awareness about proper segregation, recycling, and the hazards of improper disposal. A plastic product marked with a clear label indicating whether it is recyclable or biodegradable will help customers make informed judgments (Pilapitiya *et al.,* 2024).

Technological Innovations

Technological advances are rapidly transforming the waste management sector. Sorting technologies, including different orders, optical scanners, and artificial intelligence, are the main driving forces that lead to the efficiency and precision of plastic sorting during recycling processes. In addition, biodegradation technologies are being developed that can break down plastics faster and more effectively than natural processes. Global cooperation is fundamental to the development of uniformity in international standards for managing plastic waste, as outlined in the Basel Convention treaty, and ensuring the safe passage of plastic waste through transit nations.

Enforcement Mechanisms

Enforcement mechanisms are necessary to ensure compliance with safety standards and regulations. Regular monitoring of recycling facilities, landfills, and manufacturing units will help find non-compliant practices. Penalties for non-compliance are an effective deterrent against unsafe practices. Public reporting by industries and waste management facilities helps monitor compliance and pinpoint areas for improvement (Singh *et al.,* 2024).

Community Engagement

Community participation is key to the success of the strategies. Community-based activities (*e.g.*, cleaning, recycling, and education) are employed to encourage people to participate in plastic waste management. Collaborative work with Non-Governmental Organizations (NGOs) can lead to an increase in awareness of the campaign, thereby encouraging direct action to reduce plastic waste on the ground (Ghosh, 2019).

Overall, a complex combination of reduction at source, improved recycling infrastructure, appropriate waste management solutions, waste reduction strategies, public communication, technological innovation, global cooperation, regulatory instruments, and community participation is required for effective plastic waste management. Broadly, these methods can be applied to demonstrate how the carbon, environmental, and health and safety footprints of plastic waste can be mitigated, and as a result, contribute towards an ecologically sustainable life.

PUBLIC AWARENESS AND EDUCATION

Awareness and education play a key role in the management of plastic waste, through which behavioral changes may be triggered, and behaviorally sustainable behavior can be implemented. Awareness campaigns have been organized to educate the general public in this regard, engage them through traditional and online media, and provide solutions through television and radio, press, internet, and social networks. Influencers and celebrities can often be utilized to disseminate the message to the largest number of people. In particular, any media content that appeals to the reader, such as videos, infographics, and personal stories about the environmental impact of plastic pollution on one side and human survival on the other, can be utilized by individual social networking sites.

Educational Material

Educational materials are also important components of public awareness efforts. Educational materials (*e.g.*, infographics, brochures, videos, and documents) are required to educate people on the importance of good waste management. Schools also play a crucial role in this process by incorporating environmental education into their curriculum. From this perspective, young children should be educated about the value of waste recycling and how to minimize plastic consumption, as well as how to assist with the disposal process. Workshops and training activities can also be provided to community leaders, educators, and other stakeholders to equip them with the best practical knowledge of how waste management strategies can be implemented (Hasan, 2004).

Proper Labeling and Signage

Clear labeling and signage are valuable tools for providing consumers with information about the product choices. Plastic articles that are correctly labeled with information indicating they are recyclable or biodegradable can help consumers use the appropriate method for discarding these articles. Public signage at waste collection points is equally important. Signage indicating which bins are for recyclables, non-recyclables, and organics minimizes the potential for confusion and helps follow waste management regulations (Ding *et al.,* 2023).

Policy Implementation

The publicity campaigns should move in tandem and support the policy agenda. The policy implementation will help reduce plastic use, increase the recycling rate, and improve waste management infrastructure. Public feedback during policy review and proposed changes in policy will ensure that community concerns are taken into account and that policies remain effective over time (Schmaltz *et al.,* 2020).

Technology

Technical tools can also be used, which will play an important role in community awareness and education. Digital tools can provide information on waste disposal, recycling facilities in residential areas, and surrounding activities, which are extremely useful. For example, websites, social media groups, and forums offer additional channels for debate on the issue of plastic waste management.

Collaboration with Educational Institutions

Educational institutions play a crucial role in raising awareness and promoting action related to plastic waste management. The integration of environmental education into educational programs will help the next generation develop sustainable living behavior from an early age. Collaborations with universities to develop new technologies for manufacturing biodegradable plastic and innovative strategies for managing plastic waste will also contribute to reducing plastic waste (Liu *et al.,* 2023).

Economic Incentives

At last, economic incentives can encourage individuals and companies to take recycling more seriously. Programs that make a person eligible for compensation (on the cost of their utility bill) will encourage proper recycling of waste. Financial tax incentives for companies that use sustainable packaging (or recycling technology) will help to make the transition towards a sustainable future.

Thus, awareness and education, including a variety of activities such as media promotion, content development, clearly designated labeling, community involvement, policy lobbying, technological solutions, partnerships with educational organizations, and financial incentives, will further help achieve our goal. By integrating these components, we will ultimately lead to significant behavioral changes and the implementation of sustainable practices, which in turn will contribute to a decrease in plastic waste and pollution.

CONCLUSION

Plastic waste contamination requires a comprehensive solution that involves source reduction, improved recycling facilities, appropriate disposal options, public awareness and education, technological advancements, regulatory actions, and participatory community involvement. When properly implemented, these strategies have the potential to significantly reduce environmental pollution and health and safety impacts associated with plastic waste.

Working together across international borders is also crucial in combating this challenge. Collaborative efforts in sharing best practices, building global standards, and pursuing research and development can lead the field towards a greener future.

Ultimately, the fate of our planet will rest in our hands to manage plastic waste properly. By working together with governments, industry, communities, and the public, it is feasible to create an ecosystem in which plastic waste is significantly reduced and its harmful impacts are minimized. We have no other option but to act today to protect our environment and the health and safety of future generations.

REFERENCES

Ali, N., Khan, M.H., Ali, M., Sidra, N., Ahmad, S., Khan, A., Nabi, G., Ali, F., Bououdina, M., Kyzas, G.Z. (2024). Insight into microplastics in the aquatic ecosystem: Properties, sources, threats and mitigation strategies. *Sci. Total Environ., 913*, 169489.
[http://dx.doi.org/10.1016/j.scitotenv.2023.169489] [PMID: 38159747]

Atiwesh, G., Mikhael, A., Parrish, C.C., Banoub, J., Le, T.A.T. (2021). Environmental impact of bioplastic use: A review. *Heliyon, 7*(9), e07918.
[http://dx.doi.org/10.1016/j.heliyon.2021.e07918] [PMID: 34522811]

Barrowclough, D., Birkbeck, C. (2022). Transforming the global plastics economy: The role of economic policies in the global governance of plastic pollution. *Soc. Sci. (Basel), 11*(1), 26.
[http://dx.doi.org/10.3390/socsci11010026]

Ding, Q., Zhu, H. (2023). The key to solving plastic packaging waste lies in designing for recycling and utilizing effective recycling technology. *Polymers (Basel), 15*(6), 1485.
[http://dx.doi.org/10.3390/polym15061485] [PMID: 36987265]

Emenike, E.C., Okorie, C.J., Ojeyemi, T., Egbemhenghe, A., Iwuozor, K.O., Saliu, O.D., Okoro, H.K., Adeniyi, A.G. (2023). From oceans to dinner plates: The impact of microplastics on human health. *Heliyon, 9*(10), e20440.
[http://dx.doi.org/10.1016/j.heliyon.2023.e20440] [PMID: 37790970]

Evode, N., Qamar, S.A., Bilal, M., Barceló, D., Iqbal, H.M.N. (2021). Plastic waste and its management strategies for environmental sustainability. *Case Studies in Chemical and Environmental Engineering, 4*, 100142.
[http://dx.doi.org/10.1016/j.cscee.2021.100142]

Gao, H., Yang, B.J., Li, N., Feng, L.M., Shi, X.Y., Zhao, W.H., Liu, S.J. (2015). Bisphenol A and hormone-associated cancers: current progress and perspectives. *Medicine (Baltimore), 94*(1), e211.
[http://dx.doi.org/10.1097/MD.0000000000000211] [PMID: 25569640]

Ghosh, S. K. (2020). *Sustainable waste management: Policies and case studies.* (7th IconSWM—ISWMAW 2017, Vol. 1). Springer.
[http://dx.doi.org/10.1007/978-981-13-7071-7]

Gregory, M.R. (2009). Environmental implications of plastic debris in marine settings—entanglement, ingestion, smothering, hangers-on, hitch-hiking and alien invasions. *Philos. Trans. R. Soc. Lond. B Biol. Sci., 364*(1526), 2013-2025.
[http://dx.doi.org/10.1098/rstb.2008.0265] [PMID: 19528053]

Hasan, S.E. (2004). Public awareness is key to successful waste management. *J. Environ. Sci. Health Part A Tox. Hazard. Subst. Environ. Eng., 39*(2), 483-492.
[http://dx.doi.org/10.1081/ESE-120027539] [PMID: 15027831]

Hossain, R., Islam, M.T., Ghose, A., Sahajwalla, V. (2022). Full circle: Challenges and prospects for plastic waste management in Australia to achieve circular economy. *J. Clean. Prod., 368*, 133127.
[http://dx.doi.org/10.1016/j.jclepro.2022.133127]

Jeong, E., Lee, J.Y., Redwan, M. (2024). Animal exposure to microplastics and health effects: A review. *Emerg. Contam., 10*(4), 100369.
[http://dx.doi.org/10.1016/j.emcon.2024.100369]

Kehinde, O., Ramonu, O.J., Babaremu, K.O., Justin, L.D. (2020). Plastic wastes: environmental hazard and instrument for wealth creation in Nigeria. *Heliyon, 6*(10), e05131.
[http://dx.doi.org/10.1016/j.heliyon.2020.e05131] [PMID: 33024850]

Kibria, M.G., Masuk, N.I., Safayet, R., Nguyen, H.Q., Mourshed, M. (2023). Plastic waste: Challenges and opportunities to mitigate pollution and effective management. *Int. J. Environ. Res., 17*(1), 20.
[http://dx.doi.org/10.1007/s41742-023-00507-z] [PMID: 36711426]

Kumar, R., Verma, A., Shome, A., Sinha, R., Sinha, S., Jha, P.K., Kumar, R., Kumar, P., Shubham, N., Das,

S., Sharma, P., Vara Prasad, P.V. (2021). Impacts of plastic pollution on ecosystem services, sustainable development goals, and need to focus on circular economy and policy interventions. *Sustainability (Basel), 13*(17), 9963.
[http://dx.doi.org/10.3390/su13179963]

Kye, H., Kim, J., Ju, S., Lee, J., Lim, C., Yoon, Y. (2023). Microplastics in water systems: A review of their impacts on the environment and their potential hazards. *Heliyon, 9*(3), e14359.
[http://dx.doi.org/10.1016/j.heliyon.2023.e14359] [PMID: 36950574]

Liu, J., Hu, Z., Du, F., Tang, W., Zheng, S., Lu, S., An, L., Ding, J. (2023). Environment education: A first step in solving plastic pollution. *Front. Environ. Sci., 11*, 1130463.
[http://dx.doi.org/10.3389/fenvs.2023.1130463]

Manzoor, M.F., Tariq, T., Fatima, B., Sahar, A., Tariq, F., Munir, S., Khan, S., Nawaz Ranjha, M.M.A., Sameen, A., Zeng, X.A., Ibrahim, S.A. (2022). An insight into bisphenol A, food exposure and its adverse effects on health: A review. *Front. Nutr., 9*, 1047827.
[http://dx.doi.org/10.3389/fnut.2022.1047827] [PMID: 36407508]

Marcharla, E., Vinayagam, S., Gnanasekaran, L., Soto-Moscoso, M., Chen, W.H., Thanigaivel, S., Ganesan, S. (2024). Microplastics in marine ecosystems: A comprehensive review of biological and ecological implications and its mitigation approach using nanotechnology for the sustainable environment. *Environ. Res., 256*, 119181.
[http://dx.doi.org/10.1016/j.envres.2024.119181] [PMID: 38768884]

Matavos-Aramyan, S. (2024). Addressing the microplastic crisis: A multifaceted approach to removal and regulation. *Environ. Adv., 17*, 100579.
[http://dx.doi.org/10.1016/j.envadv.2024.100579]

Moshood, T.D., Nawanir, G., Mahmud, F., Mohamad, F., Ahmad, M.H., AbdulGhani, A. (2022). Biodegradable plastic applications towards sustainability: A recent innovations in the green product. *Cleaner Engineering and Technology, 6*, 100404.
[http://dx.doi.org/10.1016/j.clet.2022.100404]

Nøklebye, E., Adam, H.N., Roy-Basu, A., Bharat, G.K., Steindal, E.H. (2023). Plastic bans in India – Addressing the socio-economic and environmental complexities. *Environ. Sci. Policy, 139*, 219-227.
[http://dx.doi.org/10.1016/j.envsci.2022.11.005]

Nayanathara Thathsarani Pilapitiya, P.G.C., Ratnayake, A.S. (2024). The world of plastic waste: A review. *Cleaner Materials, 11*, 100220.
[http://dx.doi.org/10.1016/j.clema.2024.100220]

Sa'adu, I., Farsang, A. (2023). Plastic contamination in agricultural soils: a review. *Environ. Sci. Eur., 35*(1), 13.
[http://dx.doi.org/10.1186/s12302-023-00720-9]

vom Saal, F.S., Vandenberg, L.N. (2021). Update on the health effects of bisphenol A: Overwhelming evidence of harm. *Endocrinology, 162*(3), bqaa171.
[http://dx.doi.org/10.1210/endocr/bqaa171]

Sajjad, M., Huang, Q., Khan, S., Khan, M.A., Liu, Y., Wang, J., Lian, F., Wang, Q., Guo, G. (2022). Microplastics in the soil environment: A critical review. *Environmental Technology & Innovation, 27*, 102408.
[http://dx.doi.org/10.1016/j.eti.2022.102408]

Schmaltz, E., Melvin, E.C., Diana, Z., Gunady, E.F., Rittschof, D., Somarelli, J.A., Virdin, J., Dunphy-Daly, M.M. (2020). Plastic pollution solutions: emerging technologies to prevent and collect marine plastic pollution. *Environ. Int., 144*, 106067.
[http://dx.doi.org/10.1016/j.envint.2020.106067] [PMID: 32889484]

Shi, W., Wu, N., Zhang, Z., Liu, Y., Chen, J., Li, J. (2024). A global review on the abundance and threats of microplastics in soils to terrestrial ecosystem and human health. *Sci. Total Environ., 912*, 169469.

[http://dx.doi.org/10.1016/j.scitotenv.2023.169469] [PMID: 38154650]

Simantiris, N. (2024). Single-use plastic or paper products? A dilemma that requires societal change. *Cleaner Waste Systems, 7*, 100128.
[http://dx.doi.org/10.1016/j.clwas.2023.100128]

Singh, M., Singh, M., Singh, S.K. (2024). Tackling municipal solid waste crisis in India: Insights into cutting-edge technologies and risk assessment. *Sci. Total Environ., 917*, 170453.
[http://dx.doi.org/10.1016/j.scitotenv.2024.170453] [PMID: 38296084]

Thushari, G.G.N., Senevirathna, J.D.M. (2020). Plastic pollution in the marine environment. *Heliyon, 6*(8), e04709.
[http://dx.doi.org/10.1016/j.heliyon.2020.e04709] [PMID: 32923712]

Tumu, K., Vorst, K., Curtzwiler, G. (2023). Global plastic waste recycling and extended producer responsibility laws. *J. Environ. Manage., 348*, 119242.
[http://dx.doi.org/10.1016/j.jenvman.2023.119242] [PMID: 37832296]

Vanapalli, K.R., Sharma, H.B., Ranjan, V.P., Samal, B., Bhattacharya, J., Dubey, B.K., Goel, S. (2021). Challenges and strategies for effective plastic waste management during and post COVID-19 pandemic. *Sci. Total Environ., 750*, 141514.
[http://dx.doi.org/10.1016/j.scitotenv.2020.141514] [PMID: 32835961]

Tran, V.V., Park, D., Lee, Y.C. (2020). Indoor air pollution, related human diseases, and recent trends in the control and improvement of indoor air quality. *Int. J. Environ. Res. Public Health, 17*(8), 2927.
[http://dx.doi.org/10.3390/ijerph17082927] [PMID: 32340311]

Walker, T.R., Fequet, L. (2023). Current trends of unsustainable plastic production and micro(nano)plastic pollution. *Trends Analyt. Chem., 160*, 116984.
[http://dx.doi.org/10.1016/j.trac.2023.116984]

Williams, A.T., Rangel-Buitrago, N. (2022). The past, present, and future of plastic pollution. *Mar. Pollut. Bull., 176*, 113429.
[http://dx.doi.org/10.1016/j.marpolbul.2022.113429] [PMID: 35217417]

Yuan, Z., Nag, R., Cummins, E. (2022). Human health concerns regarding microplastics in the aquatic environment - From marine to food systems. *Sci. Total Environ., 823*, 153730.
[http://dx.doi.org/10.1016/j.scitotenv.2022.153730] [PMID: 35143789]

Ziani, K., Ioniță-Mîndrican, C.B., Mititelu, M., Neacşu, S.M., Negrei, C., Moroşan, E., Drăgănescu, D., Preda, O.T. (2023). Microplastics: a real global threat for environment and food Safety: A State of the art review. *Nutrients, 15*(3), 617.
[http://dx.doi.org/10.3390/nu15030617] [PMID: 36771324]

CHAPTER 4

Innovative Remediation Techniques for Polluted Soils: Insights from Krishi Vigyan Kendra in India

Harshad L. Jadhav[1,*], **Gaurav G. Pawar**[2], **Priyanka P. Pandey**[1] and **Sanjeev Kumar**[3]

[1] *Department of Economics and Research Centre, Karmaveer Bhaurao Patil College, New Mumbai, Maharashtra, India*

[2] *Department of Economics, Abasaheb Marathe Arts and New Commerce, Science College Rajapur, Ratnagiri, Maharashtra, India*

[3] *School of Hotel Management & Tourism, Lovely Professional University Phagwara, Punjab, India*

Abstract: The present chapter explores the types, causes, and current scenario of soil pollution in India. The role of public-sector agricultural extension and research agencies in preventing soil pollution in India has been discussed in this chapter. The public sector expenditure on agricultural extension in terms of soil and water conservation in India during the years 2009-10 to 2018-19 has been studied in the present chapter. The interventions of agricultural science centres commonly known as 'Krishi Vigyan Kendra' (KVKs) to mitigate soil pollution in the western region of Maharashtra have been studied as a case study. The study found that public-sector agricultural extension agencies like Krishi Vigyan Kendra are playing a crucial role in creating awareness among farmers about the adoption of location-specific and scientific farm practices. The adoption of such practices ensures the mitigation of soil pollution and erosion, which further stimulates agricultural sustainability.

Keywords: Agricultural extension, India, Krishi Vigyan Kendra, Soil health card, Soil pollution, Western Maharashtra.

INTRODUCTION

After independence, India emerged as an agrarian economy. The agriculture sector contributes in the form of the highest employment-generating sector in the context of the Indian economy. Being one of the crucial factors of production, soil holds paramount significance in the agricultural sector. It also serves as a

* **Corresponding author Harshad L. Jadhav**: Department of Economics and Research Centre, Karmaveer Bhaurao Patil College, New Mumbai, Maharashtra, India; E-mail: harshad.jadhav9921@gmail.com

foundation for the overall development of the country. Even in the history of mankind, most ancient civilizations emerged in the fertile soil along riverbanks. Hence, the interplay between soil health and agricultural production directly affects food security and the livelihoods of people. Soil contributes to the major ecosystem in various ways. Soil transforms organic materials into inorganic nutrients with the help of soil organisms. It helps in terms of providing water, nutrients, and physical support for plant growth and development in animal production. It regulates the CO_2, N_2O, and CH_4 emissions and supports carbon sequestration (FAO & ITPS, 2017). In recent years, soil pollution has emerged as a critical challenge in India. As the country strives towards the vision of 'Viksit Bharat', *i.e.*, a developed nation, it is important to address the issue of soil erosion and soil pollution in India. The increased industrial activities and urbanization in India have worsened the problem of soil pollution. Hence, there is an urgent need for the implementation of strict regulations to mitigate the problem. As the country has a strong network of public-sector extension agencies, the present chapter discusses the role of these agricultural extension agencies in the prevention of soil pollution.

MEANING, TYPES, AND SOURCES OF SOIL POLLUTION

Meaning and Definitions of Soil Pollution

Soil health has become a matter of global environmental issue in recent years as it impacts human health, the ecosystem, and agriculture. Being one of the parts and parcels of earth's natural resources, as well as the base of plant life, soil plays a crucial role in sustaining food grain production and maintaining biodiversity. Soil has definitely played an important role in all the progress that humans have achieved so far. However, there is always a trade-off between development and the environment. Nowadays, soil is contaminated by numerous human activities, including the disposal of industrial waste, excessive pesticide use, and urbanization. There is an urgent need to understand the meaning, nature, and causes of soil pollution, which will help in designing effective strategies to mitigate its effects and promote a healthier environment for upcoming generations. The definitions of soil pollution given by several reports are:

"Soil pollution refers to the presence of a chemical or substance out of place and/or present at a higher-than-normal concentration that has adverse effects on any non-targeted organism" (Rodríguez Eugenio *et al.,* 2018). "Soil pollution is defined as contamination of soil at higher-than-normal concentrations by waste materials of human origin that have adverse effects on human and ecosystem health" (Münzel *et al.,* 2022).

Soil pollution and soil contamination are often used interchangeably. However, according to the Intergovernmental Technical Panel on Soils (ITPS) under the Global Soil Partnership (GSP), there is a difference between these two terms. According to this report, when a chemical or any material is present in the soil at a concentration greater than what would be found naturally, it is called 'soil contamination'. Soil contamination is not always harmful. On the other hand, soil pollution is the presence of chemicals or materials that are out of place and/or present at a greater than normal concentration that negatively affects any non-targeted organism. Both intentional and inadvertent actions can lead to soil pollution. The different types of soil pollution are given in the following section.

Types of Soil Pollution

Soil pollution can be caused by a specific event or a series of events within a particular area in which contaminants are released to the soil, and the source and identity of the pollution are easily identified. This type of pollution is known as point-source pollution. This type of soil pollution is very common in urban areas. On the other hand, in Fig. (**1**), it is mentioned that soil pollution, which spreads over very wide areas, accumulates in soil, and does not have a single or easily identified source, is called diffuse pollution. It involves the transport of pollutants via the air-soil-water system (Rodríguez Eugenio *et al.,* 2018).

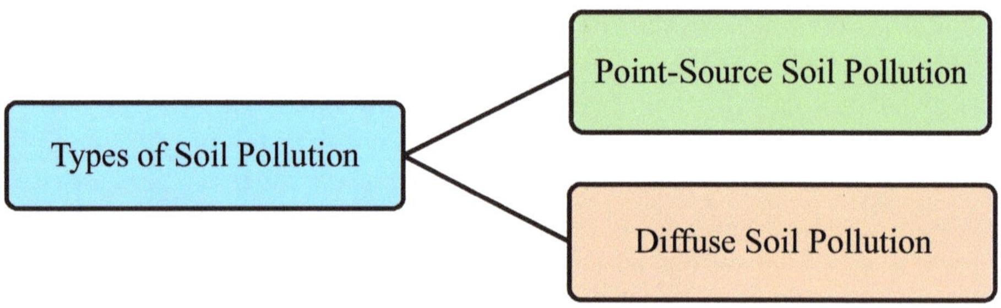

Fig. (1). Types of soil pollution. **Source:** Rodríguez Eugenio *et al.*, 2018.

Various Sources of Soil Pollution

Sources of soil pollution can be broadly classified into two types: natural (geogenic) sources and anthropogenic sources. Natural and geogenic sources of soil pollution stem from both environmental processes and geological formations. These sources include the weathering of rocks, which releases heavy metals and minerals into the soil, altering soil composition and fertility. For instance, volcanic eruptions can deposit ash and toxic substances, leading to elevated concentrations of heavy metals like lead, cadmium, and arsenic. Natural phenomena such as erosion and landslides can also expose and redistribute pollutants, while biogenic processes, such as the decomposition of organic matter, may introduce harmful pathogens or toxins. Additionally, certain natural deposits, such as those found in mining areas or high-radon zones, can contaminate soils with hazardous elements, impacting soil health and posing risks to human and ecological well-being. Understanding these natural origins is crucial for developing effective strategies for soil management and pollution remediation.

On the contrary, anthropogenic sources of soil pollution refer to human activities that result in the contamination of soil with pollutants. These sources include industrial activities, mining works, urban and transport infrastructures, waste and sewage generation and disposal, military activities, wars, and agricultural and livestock activities. Preventing anthropogenic soil pollution requires a multifaceted approach that combines effective policy, sustainable agricultural practices, and public awareness. Implementing strict regulations on industrial waste disposal and promoting the use of eco-friendly materials can significantly reduce the amount of harmful substances entering the soil.

CURRENT SCENARIO OF SOIL POLLUTION IN INDIA

With a surface area of around 3.29 million km^2 and a range of climates and landscapes, India, the world's seventh-largest country, exhibits extraordinary diversity. India's geography is as diverse as its culture, ranging from the lush hill stations of the northeast to the Thar Desert in the west, and from the majestic Himalayan Mountains in the north to the coastal plains in the south. This variation has a major impact on soil texture nationwide. While the Deccan plateau is distinguished by red and black soils that are rich in minerals but vary in fertility, the northern plains, which are mainly made up of fertile alluvial soil, sustain vast agriculture. On the other hand, the mountainous areas have loamy soils, while the coastal districts have sandy and saline soils that are appropriate for particular crops. According to the Land Use Statistics at a Glance: 2022-23 report, nearly 72021 thousand hectares of land in India come under forest cover. 27845 thousand hectares of land are utilized for non-agricultural uses. 16554 thousand hectares of land are barren and unculturable. The detailed land use pattern as per the nine-fold classification is given in Table **1**.

Table 1. Detailed land use pattern in India 2024.

Sr. No.	Classification of land	Area (in thousand hectares)
1	Forests	72021
2	Area put to non-agricultural uses	27845
3	Barren and unculturable land	16554

(Table 1) cont.....

Sr. No.	Classification of land	Area (in thousand hectares)
4	Permanent pastures and other grazing lands	10248
5	Land under Misc. tree Crops	2992
6	Culturable Wasteland	11659
7	Fallow Land Other than Current Fallows	11128
8	Current Fallow	13498
9	Net Area Sown	140705
10	Agricultural Land (5+6+7+8+9)	179982
11	Cultivated Land (8+9)	154203
12	Cropping Intensity (% of Total Cropped Area over Net Area Sown)	155.9

Source: Ministry of Agriculture and Farmers Welfare, 2024.

Natural and Geogenic Soil Pollution in India

According to the National Information System for Climate and Environment Studies portal, the majority of salt-affected soil is found in Gujarat state. The Bhal region of Gujarat state is found to be mostly affected by this kind of soil pollution. The causes of salt-affected soils in this region are inadequate irrigation systems, as well as the high rate of evaporation. As a result, these regions experience reduced agricultural productivity, decreased crop yields, and lower water table levels, ultimately affecting the livelihoods of thousands of farmers. Several states, namely Maharashtra, Karnataka, Andhra Pradesh, Telangana, Odisha, Chhattisgarh, and Madhya Pradesh, are found to be affected by water erosion. The majority of the area in the North-eastern states of India, namely, Assam and Meghalaya, comes under the hilly region. Due to increased deforestation and heavy downpours, the problem of waterlogging has emerged as a major threat to the environment. Wind erosion is another cause of soil pollution in India, particularly in arid and semi-arid areas. For instance, the Thar Desert in Rajasthan state is highly vulnerable to wind erosion, which leads to soil degradation. This type of event contributes not only to soil pollution but also reduces air quality in adjacent states, viz., Punjab, Haryana, and even Delhi.

Anthropogenic Soil Pollution in India

Anthropogenic factors are the major causes of soil pollution all over the world. India, as a developing nation, is experiencing rapid urbanization and industrialization, making it vulnerable to anthropogenic soil pollution. Impetuous use of chemical fertilizers, pesticides, industrial effluents, unregulated waste disposal methods, and insufficient legal frameworks has exacerbated the problem of anthropogenic soil pollution in India. According to the Central Pollution

Control Board, Government of India, Ankleshwar (Gujarat) is the most polluted industrial area in India, having a comprehensive environmental pollution index of 88.5. Ghaziabad in Uttar Pradesh holds the second rank among India's most critically polluted industrial areas, with a CEPI rating of 87.37, followed by Chandrapur in Maharashtra, with a CEPI rating of 83.38. The state-wise critically polluted industrial areas in India are listed in Table **2**.

Table 2. State-wise critically polluted industrial areas in India.

Sr. No.	State	Critically polluted industrial area	CEPI rating
1	Andhra Pradesh	Vishakhapatnam	70.82
2	Chhattisgarh	Kobra	83
3	Delhi	Nazafgarh drain basin	79.54
4	Gujarat	Ankleshwar	88.5
5	Haryana	Faridabad	77.07
6	Jharkhand	Dhanbad	78.63
7	Karnataka	Mangalore	73.68
8	Kerala	Greater Cochin	75.08
9	Madhya Pradesh	Indore	71.26
10	Maharashtra	Chandrapur	83.38
11	Orissa	AngulTalcer	82.09
12	Punjab	Ludhiana	81.66
13	Rajasthan	Bhiwadi	82.91
14	Tamil Nadu	Vellore (North Arcot)	81.79
15	Uttar Pradesh	Ghaziabad	87.37
16	West Bengal	Haldia	75.43

Source: Central Pollution Control Board, 2009.

Since India is considered an agrarian economy, agricultural and allied activities are another sector that contributes significantly to anthropogenic soil pollution in India. Lack of technical know-how and non-sustainable farm practices lead to improper management of farm resources. In recent years, the use of chemical fertilizers and pesticides has increased in India. According to the Ministry of Agriculture and Farmers Welfare, Government of India, the consumption of chemical pesticides in India has increased from 59168 metric tonnes in 2018-19 to 63284 metric tonnes in 2021-22. Meghalaya and Sikkim are the only states in India that are declared as organic states. The state-wise chemical pesticide consumption in India (2018-19 to 2022-23) is given in Table **3**.

Table 3. Consumption of chemical pesticides in various states during 2018-19 to 2022-23.

Sr. No.	State	Pesticide consumption (in Metric Tonnes)				
		2018-19	2019-20	2020-21	2021-22	2022-23
1	Andhra Pradesh	1689	1559	1559	1759	2001
2	Bihar	850	850	995	850	995
3	Chhattisgarh	1770	1672	1639	1740	1775
4	Goa	25	30	30	32	34
5	Gujarat	1608	1784	1573	1869	1750
6	Haryana	4015	4200	4050	4066	4066
7	Himachal Pradesh	322	881	56	279	470
8	Jharkhand	646	681	1161	450	450
9	Karnataka	1524	1568	1930	2224	1669
10	Kerala	995	656	585	532	504
11	Madhya Pradesh	540	540	691	654	598
12	Maharashtra	11746	12783	13243	13175	6814
13	Orissa	1609	1115	1158	1240	1348
14	Punjab	5543	4995	5193	5376	5130
15	Rajasthan	2290	2088	2330	2104	1865
16	Tamil Nadu	1901	2225	1834	1851	1952
17	Telangana	4894	4915	4986	4920	4920
18	Uttar Pradesh	11049	12217	11557	11688	11824
19	Uttarakhand	195	224	135	213	111
20	West Bengal	3190	3630	3630	3630	3321
21	Arunachal Pradesh	5	5	2	1.4	0
22	Assam	256	410	420	474	455
25	Mizoram	26	27	0	29	31
26	Nagaland	21	19	36	41	0
33	Jammu & Kashmir	2459	2198	3352	4086	384
24	Meghalaya	Organic State				
27	Sikkim	Organic State				
Total		**59168**	**61273**	**62146**	**63284**	**52466**

Source: Ministry of Agriculture and Farmers Welfare, 2023.

Hence, there is an urgent need for training and educating farmers for the optimum application of fertilizers and chemical pesticides. Awareness should be raised among the farmers about the adoption of organic farming practices. This can be done with the help of agricultural extension agencies in India. The following section of the present chapter discusses the agricultural extension system in India.

AGRICULTURAL EXTENSION SYSTEM IN INDIA

The current agricultural extension and research system in India that we witness today has not evolved overnight. The roots of the existing Indian agricultural extension and research system can be traced back to the pre-independence British era. Before 1871, the responsibility of agricultural research and extension was given to the Home Department of the British government. Later on, the establishment of the Imperial Council of Agricultural Research (ICAR) took place in 1929. After a prolonged struggle for freedom, India gained independence in 1947. India carried out several initiatives and Programmes to improve the agricultural extension and research system. These initiatives include the commencement of the Community Development Programme (1952), National Extension Service (1953), Centers for Region-wise Agricultural Research (1956), Intensive Agriculture District Programme (1962), and Intensive Agriculture Area Programme (1964). India carried out the Green Revolution under the visionary guidance of Dr. M. S. Swaminathan and became self-reliant in food grain production. As far as the present scenario is concerned, the Indian agricultural extension and research system has a coexistence of public and private sector agencies. However, the public sector agencies constitute a major share of the agricultural extension and research system in India. The Indian Council of Agricultural Research (ICAR) serves as an apex institution for agricultural research and extension in India (Pawar, 2023).

India has a decentralized democracy. As per Article 246 of the Indian Constitution, agriculture is a subject that belongs to the state list. Hence, the public sector expenditure on agricultural extension and research is the responsibility of the state government. Several studies found that, during the period 2009-10 to 2018-19, public sector expenditure on agricultural research and education has increased at a compound annual growth rate of 12.42% over 9 years. However, during the same period, public expenditure on soil and water conservation is found to decrease at a 9-year CAGR of -4.42%. It highlights that the public expenditure on soil and water conservation is neglected in India. Furthermore, there is a great disparity in state-wise intensity of public sector expenditure on agricultural extension and research in India.

To overcome the problem of soil pollution, the Soil Health Card (SHC) scheme was launched in India in February 2015. The prime objective of the scheme is to issue soil health cards to farmers every two years, providing a basis to address nutritional deficiencies in fertilization practices and to avoid soil pollution. The responsibility of executing the scheme is given to the agricultural extension agencies. Table **4** illustrates the state-wise progress of the soil health card scheme in the year 2023-24.

Table 4. State-wise progress of the soil health card scheme in India (2023-24).

Sr. No.	State	Total soil samples collected	Total tests completed and SHC generated	Percentage
1	Andhra Pradesh	417952	93495	22.37
2	Arunachal Pradesh	73789	18288	24.78
3	Assam	341718	260993	76.38
4	Bihar	400649	242371	60.49
5	Chhattisgarh	173751	171584	98.75
6	Goa	4904	1793	36.56
7	Gujarat	385325	238734	61.96
8	Haryana	515921	122452	23.73
9	Himachal Pradesh	74084	45918	61.98
10	Jammu & Kashmir	78369	74164	94.63
11	Jharkhand	158168	104228	65.90
12	Karnataka	181196	181156	99.98
13	Kerala	92324	28385	30.74
14	Ladakh	2342	131	5.59
15	Madhya Pradesh	803724	472085	58.74
16	Maharashtra	417766	155737	37.28
17	Manipur	385	323	83.90
18	Meghalaya	8141	5234	64.29
19	Mizoram	9822	2195	22.35
20	Nagaland	110211	103315	93.74
21	Odisha	152951	152947	100.00
22	Puducherry	2862	297	10.38
23	Punjab	120175	11458	9.53
24	Rajasthan	339573	271117	79.84
25	Sikkim	40778	29304	71.86
26	Tamil Nadu	378208	251696	66.55
27	Telangana	19069	8525	44.71

(Table 4) cont.....

Sr. No.	State	Total soil samples collected	Total tests completed and SHC generated	Percentage
28	Tripura	19495	9243	47.41
29	Uttar Pradesh	826000	826000	100.00
30	Uttarakhand	81671	41390	50.68
31	West Bengal	121727	58857	48.35

Source: Ministry of Agriculture and Farmers Welfare, 2024.

Krishi Vigyan Kendra: Concept, Origin, and Mandates

Krishi Vigyan Kendra (KVK), commonly known as 'Agricultural Science Centres', are public sector agricultural extension agencies in India which function at the grass-root level. The origins of Krishi Vigyan Kendra can be traced to the recommendations of the Kothari Commission. The third commission on education in India highlighted the need for agricultural polytechnics, which will work as specialized institutes to provide training to farmers. The first KVK in India was established in 1973 at Pondicherry. Currently, there are a total of 731 Krishi Vigyan Kendras (KVKs) functioning across the country (Waware *et al.,* 2024).

According to Pawar (2023), the mandates of KVKs in India are to conduct On-Farm Trials (OFTs) of agricultural technologies to find out the location specificity, to organize Front Line Demonstrations (FLDs) of advanced and proven agricultural technologies, to organize training Programmes for farmers, and to conduct extension activities for the rural people. Krishi Vigyan Kendras play a crucial role in the dissemination of cutting-edge agricultural technologies. Several studies have found that the scientific orientation of farmers related to improved farm practices has been significantly increased due to interventions by Krishi Vigyan Kendra (Pawar *et al.,* 2022). Krishi Vigyan Kendras are found contributing not only to agriculture but also to allied sectors like animal husbandry as well as the fishery sector (Bhave *et al.,* 2022). Hence, it is important to understand the role of agricultural extension agencies in general, and Krishi Vigyan Kendra in particular in the prevention of soil pollution and the agricultural development of India.

ROLE OF KRISHI VIGYAN KENDRA IN PREVENTION OF SOIL POLLUTION IN WESTERN REGION OF MAHARASHTRA: A CASE STUDY

Research Methodology

Selection of the Study Area

A multistage sampling technique has been used for the selection of the sample size for the present study. Maharashtra state is one of the progressive and agriculturally developed states in India. Hence, the state of Maharashtra is purposively selected for the present study. In the second stage, out of the 6 administrative divisions of Maharashtra state, the western region is randomly selected. The Western region of Maharashtra consists of a total of 5 districts, namely Satara, Sangli, Kolhapur, Solapur, and Pune. Out of these 5 districts, three districts, namely Satara, Sangli, and Kolhapur, are selected for the present case study.

Selection of Krishi Vigyan Kendra

Three Krishi Vigyan Kendras in each selected district are selected as a sample KVK for the present research. Hence, KVK Borgaon, KVK Kanchanpur, and KVK Talsande are located in Satara, Sangli, and Kolhapur districts, respectively.

Selection of Study Period

The study period selected for the present research is from 2015-16 to 2019-20.

Selection of Sample Respondents

The consolidated list of beneficiary farmers of all 3 selected KVKs has been prepared. Out of these total beneficiaries, 350 farmers have been selected as the sample for the present study. These 350 selected farmers are again distributed into 3 groups according to their respective KVKs using a proportionate random sampling method. A total of 114 beneficiaries from Satara district, 106 from Sangli, and 130 from Kolhapur district have been selected for the study. In order to carry out a comparative analysis, an equal number of non-beneficiary KVK farmers are selected. Hence, a total of 350 + 350 = 700 farmers are selected for the present study.

Adequacy of Sample Size

As per the 2011 census report, there are a total of 1727302 cultivators in the selected districts. As per Taro Yamane's formula, the adequacy of the sample size is 399.9, which is approximately equal to 400. As the sample size selected for the present study is greater than an adequate sample size, *i.e.*, 700 > 400, the present study fulfills the doctrine of adequacy of sample size.

Data Analysis and Major Findings

Composition of Micro-nutrients in Soil Under the Study Area

According to the nutrient dashboard of the soil health card portal, the soil under the jurisdiction of KVK Borgaon, Satara has 90.4% deficiency of nitrogen, 97% deficiency of Sulphur, 49.3% deficiency of iron, 34.7% deficiency of Zinc, 97% deficiency of organic carbon, 99.4% of Boron, and 7.6% deficiency of Manganese. 88.1% of the soil under the jurisdiction of KVK Borgaon, Satara, is non-saline in nature.

The soil under the jurisdiction of KVK Kanchanpur, Sangli has 96.7% deficiency of nitrogen, 49.4% deficiency of Sulphur, 37.3% deficiency of iron, and36.2% deficiency of Zinc. 90.2% of the soil under the jurisdiction of KVK Kanchanpur, Sangli, is non-saline in nature.

The soil under the jurisdiction of KVK Talsande, Kolhapur, has 58.1% deficiency in nitrogen, 32% deficiency in iron, 70.2% deficiency in zinc. 93.4% of the soil under the jurisdiction of KVK Talsande, Kolhapur, is non-saline in nature. 34.7% of soil has been found to be alkaline in nature, whereas 34.7% of soil has a neutral pH value.

As the composition of micro-nutrients is not identical throughout the study area, different agricultural techniques and practices are required. In this regard, Krishi Vigyan Kendra functions to provide location-specific guidance to farmers to help prevent soil pollution.

Soil Health Card Testing Carried Out by Sample Krishi Vigyan Kendras

Testing of soil health and distribution of soil health cards is one of the mandates of Krishi Vigyan Kendra. All Krishi Vigyan Kendras in India are provided with soil testing laboratories, where farmers can have their land soil tested at discounted rates. As per the annual progress reports of selected KVKs, the soil testing carried out by the sample KVKs during the study period is given in Table 5.

Table 5. Soil testing carried out by selected KVKs during 2015-2020.

Year	KVK Borgaon		KVK Kanchanpur		KVK Talsande	
	Soil samples tested	Soil Health Cards distributed	Soil samples tested	Soil Health Cards distributed	Soil samples tested	Soil Health Cards distributed
2015-16	460	278	422	396	405	405
2016-17	728	539	191	168	2878	2878
2017-18	463	443	1079	1054	3162	3162
2018-19	723	723	1265	1189	3245	3245
2019-20	635	635	585	562	2947	2947
Total	**3009**	**2618**	**3542**	**3369**	**12637**	**12637**

Source: Annual Progress Reports of KVK, 2015-2020.

On-farm Trails (OFTs) and Front-line Demonstrations (FLDs) Performed by Sample Krishi Vigyan Kendras

Based on the results of soil health cards, scientists carry out on-farm trials of recommended agricultural technologies and practices in order to find out the location specificity and applicability. Once the technology is found suitable for the particular region, the front-line demonstrations are carried out for the mass spread of that technology. On-farm trials and front-line demonstrations are the crucial mandates of Krishi Vigyan Kendra, which ensure that the agricultural technology is disseminated to the grassroots level, resulting in increased farm productivity. The number of on-farm trials and frontline demonstrations conducted by the sample Krishi Vigyan Kendras in the study area during the study period is given in Table **6**.

Table 6. Number of OFTs and FLDs conducted by selected KVKs 2015-2020.

Year	KVK Borgaon		KVK Kanchanpur		KVK Talsande	
	OFTs	FLDs	OFTs	FLDs	OFTs	FLDs
2015-16	2	9	7	18	7	14
2016-17	5	9	9	14	13	17
2017-18	6	11	8	13	15	18
2018-19	4	9	12	12	14	18

(Table 6) cont.....

Year	KVK Borgaon		KVK Kanchanpur		KVK Talsande	
	OFTs	**FLDs**	**OFTs**	**FLDs**	**OFTs**	**FLDs**
2019-20	6	16	6	10	10	16
Total	**23**	**54**	**42**	**67**	**59**	**83**

Source: Annual Progress Reports of KVK, 2015-2020.

Impact of Interventions of Krishi Vigyan Kendra on the Adoption of Improved Soil Management Practices in the Study Area

In order to assess the impact of interventions of KVKs on the adoption of improved soil management practices by the sample farmers, the survey method has been used. The questionnaire was designed in relation to the improved soil management practices, and the responses of the sample farmers were recorded on a scale of 1 for adopted and 0 for non-adopted. The mean score differences of each aspect have been compared using an independent sample 'z' test.

Table 7. Impact of interventions of KVK on the adoption of improved soil management practices.

Sr. No.	Improved Soil Management Practice	'z' test statistics	
		Mean diff.	**'z' value**
1	Soil testing	0.62	11.53*
2	Application of bio-fertilizers	0.36	6.383*
3	Recommended dose of chemical fertilizers	0.53	10.060*
4	Crop rotation	0.031	1.289^{NS}
5	Zero tillage	0.177	2.902*
6	Use of micro irrigation	0.07	1.363^{NS}
7	Recommended dose of pesticides	0.24	4.236*

(*Significant at 0.05 level, NS: non-significant) **Source:** Wavare *et al.,* 2024.

It is mentioned in Table 7 that there is a significant difference in the adoption scores of beneficiary farmers and non-beneficiary farmers in the case of soil testing, application of bio-fertilizers, recommended dose of chemical fertilizers, zero tillage, and recommended dose of pesticides. Hence, Krishi Vigyan Kendras are playing a crucial role in the adoption of location-specific farm practices and improved soil management practices by the farmers.

CONCLUSION

The study highlights the crucial role of agricultural extension agencies in preventing soil pollution in India. Agricultural extension agencies, in general, and Krishi Vigyan Kendras (KVKs), in particular, promote sustainable agricultural practices and protect soil health. The findings of the study indicate that anthropogenic factors are more responsible for soil pollution in India compared to natural and geogenic factors. The agricultural sector is one of the most vulnerable sectors facing the problem of soil pollution. Practices such as excessive and uncontrolled use of chemical fertilizers and pesticides, mismanagement of farm resources, and traditional irrigation techniques are the major causes of soil pollution and erosion in India. The study underscores the significance of public-sector agricultural extension and research agencies in promoting on-farm trials, frontline demonstrations, soil testing, and farmers' training in the mitigation of soil pollution. The study found that the interventions of Krishi Vigyan Kendra in the western region of Maharashtra have resulted in improved awareness and adoption of better soil management practices among the beneficiaries. Therefore, the study suggests that the government should continue to support agricultural extension and research agencies such as KVKs and scale up their funding. By doing so, they can help ensure that India's agricultural sector continues to grow while protecting its soil resources for future generations.

REFERENCES

Bhave, A.V., Wavare, A.K., Pawar, G.G., Mengal, S.G., Solkar, N.A.R. (2022). Training needs of fisherwomen in South Konkan region of Maharashtra State. *Journal of Survey in Fisheries Sciences, 9*(1), 740-745. https://www.researchgate.net/publication/376797257_Training_Needs_Of_Fisherwomen_In_South_Konkan_Region_Of_Maharashtra_State

Central Pollution Control Board, (2009). *Comprehensive Environmental Assessment Of Industrial Clusters: Ecological Impact Assessment Series (EIAS/5/2009–2010)*Ministry of Environment and Forests, Government of India. Available form: https://cpcb.nic.in/

Food and Agriculture Organization, (2017). *Global Assessment Of The Impact Of Plant Protection Products On Soil Functions And Soil Ecosystems.* Available form: https://openknowledge.fao.org/

Krishi Vigyan Kendra Borgaon, (2021). *Annual progress report of KVK Borgaon: 2015–2020.*

Krishi Vigyan Kendra Kanchanpur, (2021). *Annual progress report of KVK Kanchanpur: 2015–2020.*

Krishi Vigyan Kendra Talsande, (2021). *Annual progress report of KVK Talsande: 2015–2020.*

Ministry of Agriculture & Farmers Welfare, (2024). *Land use statistics at a glance 2022–23.* Directorate of Economics and Statistics, Government of India. https://desagri.gov.in/wp-content/uploads/2024/09/Final-file-of-LUS-2022-23-for-uploading.pdf

Ministry of Agriculture and Farmers Welfare, (2023). *Consumption of chemical pesticides in various states/UTs during 2018–19 to 2022–23.* Directorate of Plant Protection, Quarantine & Storage, Government of India. https://ppqs.gov.in/statistical-database

Ministry of Agriculture and Farmers Welfare, (2024). *Soil health card scheme progress dashboard.* https://www.soilhealth.dac.gov.in/soil-healt-dashboard

Münzel, T., Hahad, O., Daiber, A., Landrigan, P.J. (2023). Soil and water pollution and human health: what should cardiologists worry about? *Cardiovasc. Res.,* *119*(2), 440-449. [http://dx.doi.org/10.1093/cvr/cvac082] [PMID: 35772469]

National Remote Sensing Centre, Indian Space Research Organisation. (2024). *National Information System For Climate And Environment Studies.* Available form: https://nices.nrsc.gov.in/

Pawar, G.G., Wavare, A.K., Babar, M.S. (2022). Impact of interventions of Krishi Vigyan Kendra on knowledge level of soybean cultivators in Satara District of Maharashtra. *Educ. Soc.,* *1*(II), 387-394. https://www.researchgate.net/publication/365727807_Impact_of_interventions_of_Krishi_Vigyan_Kendra_on_knowledge_level_of_soybean_cultivators_in_Satara_district_of_Maharashtra

Rodríguez Eugenio, N., McLaughlin, M., Pennock, D. (2018). *Soil Pollution: A Hidden Reality*Food and Agriculture Organization of the United Nations. Available form: https://www.fao.org/3/I9183EN/i9183en.pdf

CHAPTER 5

The Impact of Environmental Pollution on Food Security and Sustainable Agriculture

Wasswa Shafik[1,2,*]

[1] *School of Digital Science Universiti Brunei Darussalam, Bandar Seri Begawan, Brunei Darussalam*

[2] *Dig Connectivity Research Laboratory (DCRLab), Kampala, Uganda*

Abstract: Widespread environmental pollution today results in contamination of soil, water, air, and, consequently, food and feed produced. This poses a significant threat to food security, defined as a state where all household members have safe access to sufficient, nutritionally adequate, and culturally acceptable food at all times, enabling them to live an active and healthy life. In the era of a growing population and concerns about environmental degradation, land, and water scarcity, the economic and health implications of increased pollution on decreased agricultural productivity, land degradation, food availability, and increased poverty deserve more attention. A detailed review of the food security implications of different types of pollution is currently missing, a gap this paper aims to fill. Several alarming statistics and actual case scenarios substantiate the urgent need for undertaking such a review. It is estimated that pesticide poisoning severely affects the health of 3 million agricultural workers in the developing world. About 200,000 deaths are caused by acute pesticide poisoning annually, leading to the loss of roughly 20 million disability-adjusted life years. Both members of the human population involved in food production and biodiversity resources are endangered. Due to the hazards associated with pesticide use, alternative biological control methods are increasingly adopted, adding a new dimension to co-occurring hazards. Ecological systems and economic interests are all interconnected, and contamination and degradation of one implies subsequent degradation of the others. In many regions, the increasing pollution of shared resources, such as rivers, streams, and pastures, is attributed to the intensification of grazing and the use of fertilizers in fodder production, as presented in the chapter. In turn, this plays a significant role in reducing the diversity of pollinator species.

Keywords: Corporate Social Responsibility, Eco-friendly Technologies, Global Impact, Green Innovation, Sustainable Business Models, Sustainable Practice.

* **Corresponding author Wasswa Shafik**: School of Digital Science Universiti Brunei Darussalam, Bandar Seri Begawan, Brunei Darussalam; Dig Connectivity Research Laboratory (DCRLab), Kampala, Uganda; E-mail: wasswashafik@ieee.org

Sanjeev Kumar, Mohammad Badruddoza Talukder & A.K. Haghi (Eds.)

INTRODUCTION

Environmental pollution refers to the contamination of materials or resources that can be directly or indirectly harmful to the environment. It is the most pressing concern for the world, resulting from human activities and products such as waste, burning fuels, and deforestation (Shafik, Tufail, Apong, *et al.,* 2024). Some of the major types of environmental pollution are air pollution, water pollution, noise pollution, soil pollution, thermal pollution, and radioactive pollution. These types of pollution have significant implications for their aggravating effects on agricultural systems (Nautiyal *et al.,* 2024). Together with agriculture, these pollutions are evolving globally and bearing consequences on environmental factors. Food security is a multifaceted concept that secures individuals, communities, and nations. It comprises three major components: the availability, accessibility, and utilization of food. Availability refers to the presence of an adequate supply of food at global, national, and household levels. Access to food means obtaining food in an appropriate manner, whether through markets, transportation, or at the homes of individuals (Sharma *et al.,* 2024).

Utilization of food is about the ability of individuals to obtain nutrients from the food they eat, enabling them to lead a vital and active life. The quantity and quality of food are necessary. Food security can be infringed upon at various levels and in numerous ways. There is no universally accepted measure of food security, making it hard to gauge exactly what the situation is like around the world. Worldwide, 950 million people are undernourished, one in every eight, and a child dies every five seconds from hunger or malnutrition-related diseases (Sandilya *et al.,* 2024). As a result of pollution, there has been a direct reduction in food production and food unavailability due to reduced quality and quantity. Thus, food security is compromised. There is degradation in the condition of flooded areas suitable for growing rice, resulting in soil erosion and nutrient depletion, which also contributes to food unavailability. Food scarcity ultimately leads to chronic illness. Approximately 2 billion people worldwide are affected. Social instability, illness, and a lack of hygiene are particularly precarious in cities, where people from different social groups, congestion, and a disposable society must coexist with the waste generated during household and industrial activities (Singh *et al.,* 2024).

Environmental pollution represents a great challenge to this generation's sustainable future. Pollution occurs in various forms, each affecting ecosystems and contributing to food supply security concerns. These include air pollution, water pollution, soil pollution, marine pollution, and noise pollution. Air pollution refers to the release of gaseous and particulate substances from natural and anthropogenic sources into the atmosphere, where these pollutants have undesirable effects; this is particularly relevant in the context of agricultural sustainability (Quayson *et al.,* 2024).

Of particular concern are emissions from the combustion of oil, natural gas, petrol, and diesel. Water pollution refers to the entry of foreign matter into sources of water that are potable, thereby degrading their quality (Kalinaki *et al.,* 2023). This can be exemplified by the presence of toxic chemicals emitted through mining, industrial activities, sewage, and chemical spills into rivers, lakes, and dams. Soil pollution is a function of the presence of chemical contamination from foreign substances and alterations to the natural characteristics of the soil, resulting in changes to its physical, chemical, and biological properties (Shafik, 2024a). For instance, nitrogen, phosphate, potassium, and other minerals that form part of soil ecology are depleted, altered, and destroyed by fertilizers, animal waste, and municipal waste. These issues are exacerbated by toxic chemicals from the petrochemical industry and acid rain, among others. Overall, ecosystems and the environment are being compromised, and if pollution continues unabated, food security will be seriously threatened. Environmental degradation, such as pollution and other constraints, increases the risk that investments in agriculture may not yield the anticipated production (Moreno *et al.,* 2024). This would harm food security outcomes.

The concept of food security revolves around the continuous availability of enough food for a healthy and active lifestyle for all members of a household. The elements that make up this definition include food availability, *i.e.*, the consistent flow of food supplies onto the market, as well as physical and economic access to food by all, which is achieved through optimal land access and production, employment, and trade relations. Food must be properly produced, *i.e.*, it must be produced through sustainable agricultural practices, while the stability of all these essential elements ensures access to enough food to provide a complete and balanced diet on a day-to-day basis (Shafik, 2025a). Staple foods and their equitable distribution are crucial to achieving food security at the household level. Food security is one of the great socio-economic challenges of our time. Ending world hunger will not only promote human development but also enhance global security and social stability. By ensuring that all people have enough to eat, we not only increase the purchasing power of the world's poorest but also reduce deaths due to malnutrition, which disproportionately affects children in developing countries (Shafik & Kalinaki, 2024). Ensuring long-term food security ultimately requires the implementation of sustainable agricultural practices. Measuring the level of food security in a community is a complex process because it is not enough to know if people have food to eat. The symptoms of so-called hidden hunger can occur even when overall food intake in terms of calories is adequate(Adebunmi Okechukwu Adewusi *et al.,* 2024). Malnutrition affects each person's ability to develop their full human potential, and it can have significant social and economic consequences.

EFFECTS OF ENVIRONMENTAL POLLUTION ON FOOD SECURITY

Increased environmental pollution has deleterious effects on various facets of the natural world, including agriculture. However, despite the gravity of the situation, the relationship between pollution and food security is largely overlooked. Food security is defined as having sufficient and consistent access to food that meets dietary needs, both physically and economically. An important part of this physical access includes the biological and environmental conditions under which the food is grown. Environmental pollution, however, impacts this directly (Abdul-Majid *et al.,* 2024).

Pollutants present in the soil, such as heavy metals, pesticides, and fertilizers, have the potential to compromise both crop and water quality. Soil is an integral part of agriculture, acting as the source of nutrients for most plants and as a medium for the growth of important crops. Contaminants in the soil, therefore, can restrict proper plant growth and soil quality; soil that is not healthy is unable to provide sustainable plant growth. Insufficient plant growth decreases yield while also lowering the nutritional quality of the crop by preventing the plant from absorbing the necessary nutrients (Sharma *et al.,* 2024). The link between pollution, the inability to grow adequate crops, and lower food quality can have a direct impact on human health and economic productivity. In practice, this has been observed in fields contaminated with environmental toxins that are unsuitable for growing food. For example, people growing crops on abandoned zinc-mine waste fields suffered from crop failure and experienced a significant increase in the infant mortality rate in the area, which is threatened by wastes that can be utilized for proper ecological conservation through organic farming (Sandilya *et al.,* 2024).

A complex system of components, including both natural and anthropogenic influences, contributes to the damage of large surfaces of fertile agricultural land. Soil and water are two of the most significant pathways through which environmental pollution can impact food security. Soil contamination is the primary cause of soil degradation due to the presence of substances that are not naturally found in the soil, and numerous reports have highlighted the significant threat to soil fertility posed by this phenomenon. Heavy metals are known for their versatility and toxic properties, and they are of particular concern in the agricultural sector. Among them, cadmium, lead, and mercury pose higher risks because they can translocate to the aerial parts of plants, ultimately entering the food chain (Shafik, Tufail, Liyanage, *et al.,* 2023). Reports from many countries indicate the presence of cadmium, lead, and mercury in various types of food consumed, with the majority of data referring to vegetables and cereals. Such heavy metals interfere with the metabolism of plants, lowering chlorophyll content, nutrient uptake, and, consequently, crop productivity (Shafik, Tufail, De Silva Liyanage, *et al.,* 2024).

Cadmium can induce oxidative stress and cellular damage in plants and the human liver. Pesticides are a significant factor contributing to the decline in soil fertility. Besides leading to direct lethality, the slow degradation time often results in the accumulation of residual pesticides in agricultural products, further contaminating the environment and endangering the health of humans who consume them. Endosulfan has been detected in soil even five years after its application, and it is detectable in various green leafy vegetables grown on soil treated with endosulfan (Shafik, Tufail, De Silva, Haji, *et al.,* 2025). At high concentrations, dietary intake of endosulfan can damage both the liver and kidneys of humans, and also has negative implications for terrestrial insects, as well as skin and eye irritation when combined with water or consumed. Similarly, the isoprothiolane residue was found in an acidic paddy field months after application, and the persistence of this pesticide can threaten wetland species (Shafik, 2024b, 2025a).

Exports may also bear the risks of pesticide residues; there is an issue of isoprothiolane residues stemming from rice. Isoprothiolane displayed declining yield in the third generation, also carrying remnants of this pesticide, which cumulatively exceeded the maximum residue levels. The persistence of pesticides in soil negatively impacts the yield and sustainability of crops. In addition, pesticides may interact with various beneficial soil organisms, such as earthworms, which contribute to the formation of soil structure, crop production, and pest control (Nautiyal *et al.,* 2024). In water, global research indicates that most pesticides are sodium-channel blocking insecticides that persist in water bodies and can have several negative implications for wetlands and waterborne species. Pesticides flushed into rivers may accumulate in the food chain. For instance, tebuconazole has been detected at concentrations and is also detected in river sediments. (Abdul-Majid *et al.,* 2024).

While land contamination has implications for the ecosystem, the impact of pollution on crop yield and nutritional quality has garnered attention due to its concomitant effects on food security. Pollutants can impact plant health and development, ultimately leading to damage to their structures and reduced growth. Substances that impair plant growth are either toxic or detrimental to plants, or they inhibit plant growth by promoting diseases that disrupt the soil-plant relationship. A combination of biological, physical, and chemical factors, or solely environmental factors, may cause environmental contamination (Sharma *et al.,* 2024). However, with increased industrial activities and related anthropogenic factors, the environment is increasingly contaminated with chemical pollutants through greenhouse gas emissions.

Various pollutants in the environment have been reported to cause multiple damage to plants. For example, soil contamination with heavy metals, metalloids, and other elements reduced the growth of sunflower crops. Assessments in various parts of the world indicate that the environmental threat to the yield and nutritional quality of a range of crops is increasing significantly due to environmental pollution. Long-term reductions in crop yield due to air pollution have been reported in rice and wheat field trials, negatively affecting food availability for millions of people in several low- and middle-income countries (Sandilya *et al.,* 2024). Moreover, loss of yield not only affects the income of farmers but also leaves farmers with an additional cost of food grains for themselves and their families. This may pose health risks to the population and a significant public health crisis at the global level. Enacting resilient policies with a scientific basis and investing in initiatives to limit the impact of pollutants on crop yields is the best way forward. An increase in crop yield will benefit resource-poor farmer communities globally. A significant portion of the crop area was contaminated with heavy metals, such as lead, cadmium, and arsenic, and their levels in crops exceeded national and international standards, leading to significant health risks for the population (Quayson *et al.,* 2024). A similar situation has been reported for crops and their human consumers in another region. Several case studies illustrated the several-fold increases in pollutant levels in the root zone soil of crops in fields affected by environmental pollution. The same patterns were observed for leafy and fruit crops grown in these contaminated fields, with implications for growers, national, and international consumers, as well as the introduction of new crops to support food security through plant genetic engineering (Sidik *et al.,* 2024).

CASE STUDIES AND EXAMPLES

Asia

Food security and health are major concerns in Mindanao, the Philippines, where public health and food safety have deteriorated significantly over the past two decades. Industrial activities and the release of pollutants have led to unsafe levels of heavy metals and pesticides, resulting in the displacement of fishing communities from their traditional lands and rendering rural populations dependent on external economic inputs. This case also highlights the urgency of addressing both food safety threats within recipient countries and pollution dynamics in sending countries (Lin & Fan, 2024). This case highlights significant issues related to eutrophication and its impacts on health, food production systems, and livelihoods. It also illustrates the probability of these problems occurring through a series of events, including the blocking of better management solutions, the promotion of local pollutant-intensive agricultural systems, and ecological and socio-economic pathways from farm to human health outcomes (Nautiyal *et al.,* 2024).

South and Southeast Asia

The long history of lead contamination in South Dongting Lake, Hunan Province, China, highlights the problems associated with neglecting pollutants related to human activity and environmental planning. This powerful case has led to significant effects of environmental pollution on human health and abnormal physiological and possibly psychological development in some affected children (Shafik, Tufail, De Silva, & Apong, 2025b). This situation is exacerbated by the development of health problems and environmental pollution resulting from climate change, which has a long history with significant policy implications that are evident to the population and contribute to poverty reduction efforts (Sandilya *et al.,* 2024)—the challenging case of the upscaling and sustainability of appropriate technological solutions for contaminated sites. Established in Manila in 1902, the factory was occasionally destroyed by fire until it was closed in 1983. The factory wastes and contaminated soil contain high levels of mercury. Residents in this area rely on locally grown vegetables as their primary food source. Paddy fields are irrigated by a creek that is contaminated with the largest amount of mercury at the dam, located 50 km from the factory (Singh *et al.,* 2024). A parallel study reveals mercury pollution of groundwater in certain areas of a coconut-growing center.

Industrial Pollution in Developing Countries

Countries undergoing rapid industrialization with weak governance institutions import hazardous wastes from developed countries through trade in both recyclables and products. There is a growing problem of toxic discharges from rapidly industrializing nations, often to the detriment of low-income communities. Increased industrialization has led to more instances of natural resource damage, with over 40% of low-income sites experiencing extensive soil contamination (Shafik, 2025b). In some instances, agricultural outputs have suffered so much that farmers have lost their entire investment or have migrated out of the region to other areas, including neighboring nations. Poor and marginalized populations, many of whom are involved in agricultural activities, are most susceptible to the combination of environmental hazards and food insecurity. There is a close relationship between industrialization and the drive for self-sufficiency, poverty reduction in the short term, and growth in degraded environments (Nautiyal *et al.*, 2024). This is due to increased demand and pressure to utilize more natural resources, a phenomenon known as growth-pollution inversion, which affects both commercial production by industry and basic subsistence needs of humans. Low-income regions also tend to have some of the highest percentages of land cultivators globally. Additionally, indigenous or local people reside in areas affected by multiple land-use activities, including industrialization, which often targets resources located on their ancestral lands (Moreno *et al.,* 2024). Industrialization, overland pollution, and greater emphasis on agriculture in low-income nations have far-reaching impacts.

Additionally, farming commodities are tied to food security efforts and projects that often support impoverished regions, particularly in sub-Saharan Africa and other developing nations, by pro-industrial development institutions. These organizations are not equipped to oversee the environmental impacts of the industry. Our paper examines how healthful nutrients in food can become contaminated by a wide range of pollutants. Regulations may be lacking, or even if they exist, they may not be enforced, so unsafe vegetables are commonly sold in local markets (Abdul-Majid *et al.,* 2024).

Pesticide Contamination in Agriculture

Modern agriculture relies heavily on synthetic chemical pesticides to control pest populations on crops. Pesticide yields are expected to increase by 10–15% over the next decade to meet human demand for food and non-food crops. While resistant plant varieties developed from genetic engineering have reduced pesticide usage by an estimated 30.4%, they are not panaceas. The overuse and misuse of synthetic chemical pesticides in agriculture have been associated with numerous negative and far-reaching consequences throughout the production, harvesting, and distribution of various types of food and feed (Singh *et al.,* 2024). A significant concern is the risk of pesticide residues on fruits, vegetables, and grains, any one of which can contribute to the ill health of millions of people who depend on agriculture for their primary diet.

This contamination also includes pesticide residues in the soil, which contribute significantly to long-term reductions in soil fertility, the slow collapse of microbial, bacterial, and lower faunal diversity, and changes in soil structure that may progressively limit the ability of much cultivated agricultural land to support sustainable, eco-friendly agriculture. Another direct result of such agricultural practices is the reduction in the overall nutrient content of vegetables and fruits, which could represent an important ecological contributor to many of the documented ontological trends in human and domesticated animal populations (Lin & Fan, 2024). It has long been argued that high levels of pesticide use are essential to maintaining world agricultural productivity; however, the cases of certain regions are symptomatic of the long-term consequences of excessive pesticide use. In each case, areas adjacent to and separate from those yielding high insecticide and related pesticide production for markets have demonstrated profound problems. These range from encephalitis to persistently revealed cancer clusters. The responsibility of pesticide manufacturers and regulatory authorities, who claim to regulate agricultural produce levels for consumer safety, is clear-cut in these examples (Santoso *et al.,* 2024). The production and use of pesticides is a significant international industry with a clear return and very clear winners and losers in terms of who gains economically by using remote sensing technologies in agriculture to increase food productivity and assurance.

MITIGATION STRATEGIES

Regulatory frameworks and policies are important steps in managing the problem. The regulatory framework can be either a command-and-control or a market-based approach. Food regulatory bodies should establish and enforce chemical residue tolerance levels in food and feed products, as well as intermediary exposure limits. On a further note, each nation must develop its regulatory framework that fulfills the criteria of its local circumstances (Abdul-Majid *et al.,* 2024). Governments also need to intervene with an international governing body to develop a uniform policy on pollutants, migrants, and metabolites. Several policies are already in place to control pollutant output from intensive agricultural systems. However, these policies have not been sufficient, as more policies are required that address complex and nonlinear interactions. Urban and peri-urban agriculture also require specific legislative and regulatory considerations (Sharma *et al.,* 2024).

Policy formulation of different strategies to address the problems is required. These strategies need to adopt an integrated approach to solving environmental degradation problems. The strategies include legal and economic remedies to control pollution and market mechanisms to mitigate and internalize the impacts of change. The strategies must also include investment decisions that maximize net value above production costs in a given set of physical and socio-economic capital constraints. The cost of degradation is a factor that drives the choice of technology. Target 16 of the SDGs aims to achieve a situation where sustainable agriculture prevails over polluting intensification, with effective regulation in place to mitigate its drawbacks. Although government policies are crucial, local engagement is necessary to maximize the policy's effects (Singh *et al.,* 2024). All these strategies need to adopt a precautionary, proactive, and integrated approach to reducing pollution in order to build environmentally sustainable intensification. Policymakers in many parts of the world encourage more proactive and preventive activities in agriculture to reduce pollution. Agriculture in areas where public goods are at risk will have to undergo significant structural and knowledge-based transformations to deliver these goods. Sustainable agriculture requires public policies and support that integrate socio-economic and environmental dimensions, as well as site-specific solutions. In addition, productivity needs to be sustainable, and development must be participatory to encourage social inclusion (Abdul-Majid *et al.,* 2024). Future agriculture will need new forms of technology and innovation to produce more with fewer inputs, more efficient crop cycles, and less waste and damage.

Public awareness and stakeholder involvement. The primary challenge for policymakers is to engage the local population in the decision-making process regarding mitigation measures. Agricultural activities and the food chain may be involved in various types of pollution. Effluents from various industries, intensive aquaculture, and agriculture are potential sources of pollution (Shafik, Tufail, De Silva, & Apong, 2025a). Uncontrolled wastewater discharges and improper waste disposal can lead to environmental contamination and contamination of the food chain. The drinking water supply may also be affected by the contamination of receiving water bodies with hazardous chemicals and heavy metals, or the formation of harmful algal blooms (Sandilya *et al.,* 2024). Natural resource management policy often lacks integration of water resource management through the digitization of farming, as presented with four top components.

Water and food safety regulations, standards, and guidelines. The policy approach towards implementing regulations and standards can lead to inconsistencies in monitoring and law enforcement. Weak coordination among agencies sometimes results in the incomplete implementation of the existing legal framework. Ineffective monitoring of industrial operations may lead to non-compliance or create opportunities for quick, short-term financial gain at the expense of the environment and long-term human health. Regulatory issues often impact the ability to ensure the safety of meat and fish (Quayson *et al.,* 2024). Generally, regulations or protocols are put in place, but they are not consistently enforced or are enforced sporadically. For example, sturgeon breeding farms feed their sturgeons with uncontrolled amounts of antibiotics. More funding is spent on controlling antibiotics in meat plants than on monitoring their use on farms. In some cases, subsidies and bonuses are provided only if certain parts of the regulations are complied with, creating further issues (Lin & Fan, 2024).

Sustainable agricultural practices are one way to mitigate the impacts of various environmental pollution sources and their harmful effects on food security. This section discusses how various agricultural techniques can be applied to reduce environmental pollution, particularly water pollution, and consequently increase food production. Sustainable agricultural practices involve adopting good management and soil conservation techniques. Farmers utilize such techniques in agriculture to enhance soil fertility, thus reducing operational costs by minimizing input application. It enables farmers to avoid declines in crop quality and quantity, as it is a viable long-term solution. Some of the sustainable agricultural practices include organic farming, no-tillage, crop rotation, and integrated pest management (Abdul-Majid *et al.,* 2024). Healthy soil contributes to healthy crops, which, in turn, support healthy people. Erosion of soil also has a certain effect on air quality and biodiversity.

In addition to pollution amelioration, the application of some of those techniques results in positive changes in soil properties such as pH, organic matter content, and overall soil productivity. Organic farming reduces environmental pollution, conserves soil moisture, keeps the soil cool, increases its depth, and provides free nutrients through biological nitrogen fixation (Shafik *et al.,* 2025). Crop rotation reduces the dependence of crops on chemical pesticides and fertilizers, thereby providing safe food that is free from pesticide contamination. Such farming attracts consumer interest and drives market demand. Sustainable agricultural practices also encompass the entire farm, considering both short-term and long-term economic aspects. The technology utilized in sustainable agriculture should be in harmony with the farm and environment (Quayson *et al.,* 2024). The farmer thus gains satisfaction from the work accomplished and is assured that there will be more prosperous and productive years ahead. The perspectives should be harmonized, emphasizing the system's development over time while its subsystems are complementary. The farmer must be empowered through recruitment, education, and involvement in zones of interest related to sustainable agriculture. The application of science and technologies could bring about development and achievement in this sector. Opportunities and constraints peculiar to this approach in Africa are discussed, with case studies in Kenya and Ethiopia (Lin & Fan, 2024). The models demonstrate significant yield increases and reduced workload.

FUTURE PERSPECTIVES AND RESEARCH DIRECTIONS

This section thoughtfully examines the potential pathways for deeper research in the field of food security and environmental pollution. It identifies new issues, such as the importance of capturing the role of nutrition, new practices, and changes foreseen in the SDS. It suggests that interdisciplinary research combining social and natural science methods will be well-placed to address these emerging scientific and policy issues. It provides a roadmap for engagement with policymakers, stakeholders, and communities in local case study process workshops (Quayson *et al.,* 2024). Downstream effects on human health and the environment under different pollution scenarios require heightened interdisciplinary research approaches that consider combined exposure to multiple chemicals from various sources. Adequate technology and practice-based response systems must be identified, and any potential information asymmetries must be corrected to minimize adverse effects on human health and the environment. Another potential tangent for developing the research undertaken in the SDS is the focus on innovations needed if novel food systems are indeed viable, such as the use of nanotechnology or GM technologies to mitigate pollution levels (Sandilya *et al.,* 2024)

To develop effective future strategies, further research is necessary to identify the necessary changes in practices or policies. This will involve research to understand adaptive capacity, resilience, and the trade-offs that changes in environmental pollution levels might incur (Shafik, Tufail, Namoun, *et al.,* 2023). Suppose environmental levels of pollution are increasing to such an extent that policy measures are becoming ineffective, or an overarching global change issue constrains the implementation of policies. In that case, it may be worthwhile to investigate a set of parallel investigative tools, such as case study analysis, that enhance engagement with public and stakeholder groups to design new economic pathways (Sandilya *et al.,* 2024). Policymakers, scientists, and agricultural stakeholders must collaborate to implement a comprehensive action plan that prevents or mitigates the effects of environmental pollution. Local case studies will be essential in transferring this research. Policymakers and scientists need to focus on interventions aimed at improving the resilience and adaptive capacity of farmers and their local communities, as this may currently be the only realistic route to protect them from the impacts. The externalization of negative environmental effects is likely to continue under the BAU scenario, raising concerns over food production as a dominant end of land use (Lin & Fan, 2024; Moreno *et al.,* 2024).

REFERENCES

Abdul-Majid, M., Zahari, S. A., Othman, N., Nadzri, S. (2024). Influence of technology adoption on farmers' well-being: Systematic literature review and bibliometric analysis. *Heliyon* (2) [http://dx.doi.org/10.1016/j.heliyon.2024.e24316]

Temidayo Olorunsogo, , Temidayo Olorunsogo, , Ejuma Adaga, , Olorunsogo, T., Adaga, E., Daraojimba, D.O. (2024). AI in precision agriculture: A review of technologies for sustainable farming practices. *World Journal of Advanced Research and Reviews, 1*(1), 2276-2285. [http://dx.doi.org/10.30574/wjarr.2024.21.1.0314]

Kalinaki, K., Shafik, W., Gutu, T. J. L., Malik, O. A. (2023). Computer vision and machine learning for smart farming and agriculture practices. *Emerging Technologies In Agriculture*IGI Global. [http://dx.doi.org/10.4018/978-1-6684-8516-3.ch005]

Lin, J., Fan, Y. (2024). Seeking sustainable performance through organizational resilience: Examining the role of supply chain integration and digital technology usage. *Technol. Forecast. Soc. Change, 198*, 123026. [http://dx.doi.org/10.1016/j.techfore.2023.123026]

Moreno, J. C., Berenguel, M., Donaire, J. G., Rodríguez, F., Sánchez-Molina, J. A., Guzmán, J. L., Giagnocavo, C. L. (2024). A pending task for the digitalisation of agriculture: A general framework for technologies classification in agriculture. *Agricultural Systems* [http://dx.doi.org/10.1016/j.agsy.2023.103794]

Nautiyal, C.T., Nautiyal, P., Papnai, G., Mittal, H., Agrawal, K., Shivani, , Vishesh, , Nandini, R. (2024). Importance of smart agriculture and use of artificial intelligence in shaping the future of agriculture. *J. Sci. Res. Rep., 30*(3), 129-138. [http://dx.doi.org/10.9734/jsrr/2024/v30i31864]

Quayson, M., Bai, C., Sarkis, J., Hossin, M. A. (2024). Evaluating barriers to blockchain technology for sustainable agricultural supply chain: A fuzzy hierarchical group DEMATEL approach. *Operations Management Research.* [http://dx.doi.org/10.1007/s12063-024-00443-x]

Sandilya, D., Bharali, C., Ringku, A., Sharma, B. (2024). Utilizing greenhouse technology towards sustainable agriculture using IoT "TechFarm." In *Lecture notes in electrical engineering* (Vol. 1061). Springer. [http://dx.doi.org/10.1007/978-981-99-4362-3_34]

Santoso, H., Yusuf, M.A., Rahutomo, S., Madiyuanto, , Winarna, (2024). A cutting-edge precision agriculture technology to support the sustainable oil palm industry. *IOP Conf. Ser. Earth Environ. Sci., 1308*(1), 012053. [http://dx.doi.org/10.1088/1755-1315/1308/1/012053]

Shafik, W. (2024). Barriers to implementing computational intelligence-based agriculture system. In M. G. Sumithra, M. Sathyamoorthy, M. Manikandan, R. K. Dhanaraj, & M. Ouaissa (Eds.), *Computational intelligence in internet of agricultural things* (pp. 193–219). Springer. [http://dx.doi.org/10.1007/978-3-031-67450-1_8]

Shafik, W. (2025). Sustainable Agriculture and Diet for Urban Development.*Sustainable Smart Cities and the Future of Urban Development.* IGI Global Scientific Publishing. a

Shafik, W., Kalinaki, K. (2024). Societal and ethical implications of technology-enhanced agriculture and healthcare: An African context. *2024 Ist-Africa Conference (Ist-Africa),* IEEE.1-11. [http://dx.doi.org/10.23919/IST-Africa63983.2024.10569306]

Shafik, W., Tufail, A., Apong, R. A. A. H. M., De Silva, L. C. (2024). Internet of Things for smart agricultural practices. *Internet Of Things Applications And Technology* Auerbach Publications. [http://dx.doi.org/10.1201/9781003458401-13]

Shafik, W., Tufail, A., De Silva, L.C., Apong, R.A.A.H.M. (2025). A lightweight deep learning model for multi-plant biotic stress classification and detection for sustainable agriculture. *Sci. Rep., 15*(1), 12195. a [http://dx.doi.org/10.1038/s41598-025-90487-1] [PMID: 40204810]

Shafik, W., Tufail, A., De Silva Liyanage, C., Apong, R.A.A.H.M. (2024). Using transfer learning-based plant disease classification and detection for sustainable agriculture. *BMC Plant Biol., 24*(1), 136. [http://dx.doi.org/10.1186/s12870-024-04825-y] [PMID: 38408925]

Shafik, W., Tufail, A., Liyanage De Silva, C., Awg Haji Mohd Apong, R.A. (2025). A novel hybrid inception-xception convolutional neural network for efficient plant disease classification and detection. *Sci. Rep., 15*(1), 3936. [http://dx.doi.org/10.1038/s41598-024-82857-y] [PMID: 39890849]

Shafik, W., Tufail, A., Namoun, A., De Silva, L.C., Apong, R.A.A.H.M. (2023). A Systematic Literature Review on Plant Disease Detection: Motivations, Classification Techniques, Datasets, Challenges, and Future Trends. *IEEE Access, 11*, 59174-59203.

[http://dx.doi.org/10.1109/ACCESS.2023.3284760]

Sharma, A., Prakash, A., Bhambota, S., Kumar, S. (2024). Investigations of precision agriculture technologies with application to developing countries. *Environment, Development And Sustainability, 27*(7), 15135-15171. Available form: https://doi.org/10.1007/s10668-024-04572-y [http://dx.doi.org/10.1007/s10668-024-04572-y]

Sidik, M. A. M., Leman, A. M. bin, Feriyanto, D., Abdulmalik, S. S., Zakaria, S. (2024). Green technology for sustainable agriculture: Bio-fertilizer production from municipal waste to preserve the environment. *International Journal Of Innovation In Mechanical Engineering And Advanced Materials, 5*(3), 118. [http://dx.doi.org/10.22441/ijimeam.v5i3.23743]

Singh, M., Sachchan, T. K., Sabharwal, P. K., Singh, R. (2024). Smart and sustainable food production technologies. In: Thakur, M., (Ed.), *Sustainable Food Systems* Springer, Cham. Available form: https://doi.org/10.1007/978-3-031-46046-3_1 [http://dx.doi.org/10.1007/978-3-031-46046-3_1]

CHAPTER 6

Harnessing Innovation in Air Pollution Control and Prevention Technologies: Current Trends and Future Prospects

Parambir Singh[1], **Amrik Singh**[2,*] and **Amanjeet Bhalla**[1]

[1] *Chitkara College of Hospitality Management, Chitkara University, Rajpura, Punjab, India*

[2] *School of Hotel Management and Tourism, Lovely Professional University, Jalandhar, Punjab, India*

Abstract: Air pollution is indeed one of the most significant environmental health issues today. It results from the combined contribution of effluent discharges from industries, vehicular exhausts, agricultural activities, and natural causes. With effective air pollution control technologies and preventive measures, environmental and public health impacts will be minimized. This chapter examines and discusses various technologies for managing and reducing air pollution, including their effectiveness, implementation modalities, and potential future improvements. This chapter examines the current trends and prospects in air pollution control, as well as the prevention technologies.

Keywords: Air Pollution Control, Emission Reduction, Filtration Technologies, Renewable Energy, Environmental Policy, Sustainable Urban Planning.

INTRODUCTION

Filtration systems, such as HEPA filters, baghouse filters, and activated carbon filters, are used to remove particulate matter and other harmful gases. Electrostatic precipitators (ESPs) utilize electrical charges to capture fine particles from industrial emissions. They are regarded as one of the most effective methods of treatment for high efficiency and low cost. Various scrubbers, including both wet and dry types, serve a significant function in removing harmful gases from the atmosphere, particularly sulphur dioxide (SO_2) and nitrogen oxides (NO_X). The primary function of catalytic converters is to reduce the emission of harmful particulates from vehicles by converting harmful gases into less toxic forms. Some examples of advanced emerging green technologies include photocatalytic

* **Corresponding author Amrik Singh:** School of Hotel Management and Tourism, Lovely Professional University, Jalandhar, Punjab, India; E-mail: amrikmhm@gmail.com

Sanjeev Kumar, Mohammad Badruddoza Talukder & A.K. Haghi (Eds.)

oxidation, biofilters, and Carbon Capture and Storage (CCS), which can lead to improved long-term air quality control. Finally, there are eco-friendly alternatives to conventional fuel-based vehicles, like electric and hydrogen-powered vehicles. Policy interventions, including the Clean Air Act, European Union Air Quality Directives, and Emission Trading Systems (ETS), regulate emissions and promote the sustainable management of industries. That way, the reduced emissions will help the environment and public health in future generations. The technology and regulatory reduction model demonstrates dual effectiveness in addressing air pollution and protecting public health and environmental resources in the future.

Here is an excerpt describing the growing threats posed by air pollution to society, ecosystems, and the climate. Other kinds of environmental issues resulting from increasing pollutant levels in the atmosphere include global warming, acid rain, and respiratory diseases (Marco *et al.,* 2019). The various sources of air pollution include industrial waste, vehicle exhaust, agricultural activities, and natural phenomena such as wildfires and volcanic eruptions. This is why a strategy combining regulatory measures and technological innovations is needed to curb the health and environmental impacts of air pollution. Common air pollutants include Particulate Matter (PM), nitrogen oxides (NO_X), sulphur oxides (SO_X), Volatile Organic Compounds (VOCs), and greenhouse gases such as carbon dioxide (CO_2) and methane (CH_4) (Suh *et al*, 2000). These pollutants arise from numerous anthropogenic and natural sources. Industrialisation emits significant amounts of NO_X and SO_X, which contribute to the formation of smog and acid rain. Transportation is another major source of VOCs and PM, and other sources contribute to the worsening air quality in cities (Fenger, 1999). The greenhouse gases CO2 and CH4 are sources of climate change, resulting in increased temperatures worldwide and extreme weather events (Montzka *et al.,* 2011).

Health issues from air pollution are extensive. Fine particulate matter (PM2.5) is a cause of respiratory and cardiovascular diseases, lung cancer, and premature death (Neas, 2000). Children, the elderly, and those with pre-existing conditions are particularly at risk. Long-term exposure tends to inhibit children's lung development, leading to Chronic Obstructive Pulmonary Disease (COPD) in adults (Nadia *et al*, 2015). The economic burden of such illnesses would also be significant, as they would increase spending on healthcare and reduce productivity in the workforce.

Governments and organizations have aimed to reduce emissions by setting regulatory frameworks to protect air quality. Policies such as the Clean Air Act in the United States, the Air Quality Directives of the European Union, and the Paris Agreement have established excellent standards for emission restrictions while promoting sustainable industrial practices (Kuklinska *et al.,* 2015). These regulations facilitate the development and application of cleaner technologies, emission trading schemes, and stricter motor vehicle emission requirements.

Major technological developments play a role in reducing air pollution. High-tech filtration processes, such as HEPA filters and baghouse filters, do an excellent job of entrapping airborne particles in industrial applications. Electrostatic precipitators work on the principle of applying an electrical charge to the movement of fine particulates out of the emissions stream, thereby improving air quality in power generation facilities and industrial plants (Krigmont, 2023). The scrubbers, whether wet or dry, neutralize gases like SO_2 and NO_X, preventing the formation of acid rain. Owing to catalytic converters, which minimize noxious emissions by transforming toxic gases into less harmful compounds, vehicular emissions have been reduced (Ghulomjonovna, 2025).

Some emerging green technologies may provide sustainable solutions to air pollution prevention. Carbon capture and storage technology ensures that CO_2 emissions from industrial sources are captured and stored, rather than released into the atmosphere. The new generation of electric and hydrogen vehicles, which are free of tailpipe emissions, will combat urban air pollution (Sarfraz, 2021). Innovations in photocatalytic oxidation and biofilters are also being considered for industrial applications to degrade air pollutants.

In summary, air pollution poses an unprecedented challenge and requires a thoroughly orchestrated effort, ranging from policy measures to technological advancements. Emission reductions form the bedrock of regulatory frameworks designed to curb pollution in the real world, and thus, technology provides a means to these ends. There is a need for a global consensus on sustainable operations, the adoption of clean energy, and a comprehensive focus on research in pollution control technologies to improve air quality and safeguard public health for future generations.

Health and Environmental Impacts of Air Pollution

Air pollution is a worldwide concern, given its grave effects on human health and the environment. The resources causing air pollution include industrial emissions, transport, and agriculture, which release Particulate Matter (PM), Nitrogen Oxides (NOx), sulphur oxides (SOx), Volatile Organic Compounds (VOCs), and heavy metals into the atmosphere. These pollutants are responsible for a variety of health complications and environmental deterioration; thus, air pollution is a critical situation that requires immediate action.

Health Impacts

Numerous respiratory and cardiovascular diseases are caused by exposure to air pollution. Fine particulate matter (PM2.5) is particularly hazardous; it can deposit deep within the airways and even migrate into the bloodstream, leading to diseases such as asthma, Chronic Obstructive Pulmonary Disease (COPD), lung cancer, and cardiovascular diseases (Franklin *et al.,* 2019). Hesterberg *et al.* (2021) state that approximately 7 million people die prematurely each year due to air pollution-related illnesses. In addition, prolonged exposure to pollutants such as nitrogen dioxide (NO_2) has been linked to an increased risk of air pollution-related illnesses, reduced lung function, and an increased risk of strokes (Burnett *et al.,* 2018). Children and the elderly are particularly susceptible to these health impacts; for instance, prenatal pollution exposure has been reported to be associated with low birth weight and developmental disorders (Perera, 2017).

Environmental Effects

Moreover, air pollution causes significant environmental impacts, such as acid rain, which is a product of the reaction between sulfur dioxide (SO_2) and nitrogen oxides (NOx) when they are intermixed with water in the atmosphere. This has led to the acidification of lakes, rivers, and soils, negatively affecting aquatic life and overall agricultural production (Grennfelt *et al.,* 2020). Ground-level ozone (O_3) is another type of pollutant that reduces agricultural productivity, causing crop damage and lower yields, which in turn lead to reduced food security (Mills *et al.,* 2018). The forests are not spared; they also suffer from having pollution-associated effects because those nutrients are lost, making trees susceptible to diseases and climate change (Matyssek *et al.,* 2010). Besides humans, air pollution in high amounts leads to the development of smog, which reduces visibility and is detrimental to the breathing capabilities of inhabitants. The areas most affected are the industrial sites and cities with heavy vehicular emissions, and megacities like Delhi, Beijing, and Los Angeles often experience dangerous levels of air quality (Gurjar *et al.,* 2010). Greenhouse gases, such as carbon dioxide (CO_2) and methane (CH_4), are emitted from industrial activities and contribute to global warming (IPCC, 2021). These, in turn, further exacerbate climate change, which ultimately leads to extreme weather phenomena.

Sources of Air Pollution

Air pollution has varied origins, whether from natural or anthropogenic sources, and has a significant impact on the environment and public health. Natural sources, including wildfires, volcanic eruptions, and dust storms, are among the major causes of deteriorating air quality worldwide (Gurjar *et al.,* 2010). The major anthropogenic sources of air pollution include industrial activity, vehicular exhaust, the use of firewood, agricultural activities, and domestic energy use.

Industrial Emissions

The provision of air-polluting substances, including sulphur dioxide (SO_2), nitrogen oxides (NOx), carbon monoxide (CO), and other particulate matter (PM), occurs through various industrial activities (Lelieveld *et al.,* 2015). The combustion of fossil fuels in power plants and manufacturing industries is one of the main contributors to air pollution, alongside coal-fired power plants, which emit the most SO_2 and fine particulate matter (Liu *et al.,* 2007). Respiratory diseases and acid rain, linked to industrial emissions, also have negative consequences for ecosystems and human health.

Vehicular Exhaust

Transportation is another significant contributor to air pollution in many cities, particularly in urban areas. This pollution comes in the form of emissions such as those from the combustion of gasoline and diesel fuels in vehicles, including carbon monoxide, nitrogen oxides, volatile organic compounds, as well as fine particulate matter (Goshua *et al*, 2022). The combustion of diesel engines, in particular, results in the emission of black carbon, which is among the PM2.5 counts that pose respiratory hazards and also contribute to climate change (Shindell *et al.,* 2016). Many cities around the world are already experiencing deteriorating air quality due to urbanization and the increasing number of vehicles.

Agricultural Activities

Air pollution results from the application of chemical fertilisers, pesticides, and gases released from livestock. Ammonia released from fertilisers and animal waste undergoes a reaction with other air pollutants to give rise to fine particulate matter (Wyer *et al*, 2022). Additionally, livestock digestion produces methane (CH_4), a significant greenhouse gas that contributes to global warming. Deforestation, which occurs on a large scale for agricultural purposes, reduces the planet's ability to absorb CO_2 emissions.

Biomass Burning

Burning wood and other organic substances releases carbon monoxide, smoke, and particulate matter into the atmosphere. Open burning of agricultural waste and wildfires have been significant contributors to regional air pollution and haze formation, particularly in developing countries (Crutzen & Andreae, 1990). The burning of biomass is known to deteriorate air quality and contribute to climate change by emitting black carbon, which absorbs heat in the environment and thus accelerates global warming.

Household Energy Use

Indoor and outdoor air pollution sources in low-income regions primarily include biomass (such as wood, dung, and charcoal) combustion products and coal burning for cooking and heating. The incomplete combustion of these fuels emits carbon monoxide, black carbon, and PM2.5, among other toxic byproducts, posing significant health risks, particularly for women and children (Bonjour *et al.,* 2013). Indoor air pollution from household energy is attributed to up to millions of premature deaths from respiratory diseases and heart conditions annually (Goshua *et al.,* 2022).

Air Pollution Control Technologies

Air Pollution Control Technologies: These technologies mitigate the adverse effects of airborne pollutants on human health and the environment. Air pollution control technologies address Particulate Matter (PM), Nitrogen Oxides (NOx), Sulfur Oxides (SOx), Volatile Organic Compounds (VOCs), carbon dioxide (CO_2), and other pollutants. Such improvements in air pollution control have led to the development of filtration systems and other advancements, including those with AI and nanomaterial integration (Fayyazbakhsh *et al.,* 2022).

Filtration Systems

These filters are widely used in industrial facilities, residential air purifiers, and vehicular applications to promote indoor and outdoor air quality. HEPA filters can remove as much as 99.97 percent of airborne particles as small as 0.3 microns, making them an indispensable device in air pollution mitigation (Adepoju *et al.,* 2023).

Electrostatic Precipitators (ESP)

Electrostatic precipitators are devices that use electrostatic energy to remove particulate matter from industrial emissions. They are widely used in coal-fired power plants and the cement industry, where they are used to collect fine particles by making charges on them, which are then collected on opposite-charged plates. Reduction in particulate emissions by ESPs has been reported to be above 99%, thus making this technology vital in industrial air pollution control (Molchanov *et al.,* 2023).

Scrubber Systems

Scrubbers are air pollution control devices that capture particulates or gases from emissions through contact with a liquid, a gas, or a combination of both. Wet scrubbers use water or alkaline solutions for scrubbing to remove SOx and other acidic gases, whereas dry scrubbers use sorbents to neutralize pollutants before they can enter the air (Vig *et al.,* 2023). These systems are widely used in power plants, waste incineration plants, and chemical plants.

Catalytic Converters

Catalytic converters are essential accessories in automobiles that effectively control emissions from vehicles. These convert highly poisonous gases, such as CO, NOx, and hydrocarbons, into less poisonous forms, such as CO2 and water (Ghulomjonovna, 2025). Platinum, palladium, and rhodium are metals used in catalytic reactions to convert poisonous gases into less poisonous gases. Urban air quality is significantly improved by this technology in terms of transportation and pollution.

Biofiltration

Biofiltration employs microorganisms to degrade pollutants in the environment, especially VOCs emitted as waste from various industries (Delhoménie & Heitz, 2005). Biofilters consist of bioactive media that house microbial communities, which degrade pollutants into harmless compounds. It is commonly used in wastewater treatment plants, compost facilities, and industrial facilities that emit organic waste.

Carbon Capture and Storage (CCS)

CCS technology involves capturing CO_2 emissions produced primarily by industrial sources and power plants, thereby preventing their release into the atmosphere. The captured CO_2 can be either stored underground in geological formations or sold for use in industrial applications, such as enhanced oil recovery (Bui *et al.,* 2018). CCS is identified as one of the key technologies to aid in climate change mitigation through reduced greenhouse gas emissions.

Emerging Technologies

Photocatalytic Oxidation: Inexpensive and solar-driven purification methods are employed using light-driven catalysts, such as titanium dioxide (TiO_2), to oxidize pollutants into harmless substances (Li *et al.,* 2022).

Nanotechnology

Novel materials, like the graphene-based filter, improve fine particulate matter and toxic gas removal (Liu *et al.,* 2023).

Artificial Intelligence (AI) in Air Quality Management: AI systems analyze air quality data to optimize pollution control measures and predict pollution trends for effective management (Ye *et al.,* 2020).

Green Roofs and Urban Vegetation: By filtering a variety of air contaminants, plants act to decrease concentrations of CO_2 and thereby improve urban air quality through the uptake of NOx and PM (Perini & Magliocco, 2014).

Prevention Strategies

The primary approaches for mitigating air pollution are source reduction and promoting sustainable practices through policy interventions, technological innovations, and behavioral changes. These approaches require concerted effort from the government, practitioners in various industries, and individuals in efforts to improve air quality and protect public health.

Policy and Regulations

Countries worldwide have established air quality laws that regulate emissions from various industrial and transportation sectors. Regulatory frameworks such as the Clean Air Act (USA) and the Air Quality Directive of the European Union lay down stringent limits concerning pollutants such as particulate matter (PM), nitrogen oxides (NOx), sulphur dioxide (SO_2), and Volatile Organic Compounds (VOCs) (Kuklinska *et al,* 2015). These include enforcing emission limits, encouraging cleaner production technologies, and giving other incentives for industries to adopt pollution control methods (Schnell & Prather, 2017). Pricing carbon and emission trading systems also helps reduce greenhouse gas emissions from industries.

Renewable Energy Transition

Transitioning from fossil fuels to renewable sources such as solar, wind, and hydropower significantly reduces air pollution. The adoption of renewables reduces inputs of CO_2, NOx, and SO_2, which contribute to air pollution and global climate change (Jacobson, 2017). Policies that promote Electric Vehicles (EVs) and hydrogen fuel cells help mitigate pollution caused by transport (Waseem *et al.,* 2023). Various governments provide subsidies and incentives to increase investments in greener energy technologies.

Urban Planning and Green Infrastructure

Urban planning plays a crucial role in mitigating air pollution. The increase in green spaces (*e.g*., parks, urban forests) would help absorb pollutants and improve air quality (Monteiro *et al.,* 2020). Sustainable urban design would integrate improvements to public transport, pedestrian and cycling infrastructure, and low-emission zones to curb vehicular emissions (Wang *et al.,* 2022). Smart city initiatives involve the installation of air quality monitoring systems to watch pollution levels and perform responsive actions.

Public Awareness and Behavioural Changes

The active engagement of the public is crucial to the success of efforts to prevent air pollution. Awareness campaigns encourage the public and enterprises to adopt environmentally compatible behaviors, such as using public transport, conserving energy, and purchasing green products (Thompson, 2024). Informing communities about the health effects of air pollution motivates them to adopt measures for reducing activities that contribute to air pollution, such as waste burning and the use of energy-efficient appliances. Corporate responsibility programs also promote sustainable business practices that contribute to pollution reduction.

Case Studies

Air Pollution Control Strategies of China

China enforces strict regulations concerning industrial emissions and incentivizes the use of electric vehicles. Measures such as pollution taxes and improvements in public transportation have significantly improved air quality in major cities like Beijing.

European Low Emission Zones

Several cities in Europe have designated low-emission zones, which restrict the entry of high-polluting vehicles. This has resulted in a significant reduction in NOx and PM emissions.

Odd-even Rule of Delhi

A unique traffic regulation, known as the odd-even rule, was implemented in the capital city of India to reduce congestion and emissions on the streets. While the effects were temporary, the intervention demonstrated that policy-driven measures could restore some degree of control over air pollution.

Future Prospects and Challenges

Advancements in air pollution control technologies have significantly improved air quality in many regions. However, widespread implementation faces challenges such as high costs, inadequate infrastructure, and political constraints. Many developing countries struggle with the funding and technical expertise required for the large-scale adoption of clean technologies, while industries often resist strict regulations due to economic concerns (Kuklinska *et al.,* 2015).

Future efforts should focus on:

- **Integrating smart air quality monitoring systems** that utilize artificial intelligence (AI) and the Internet of Things (IoT) to provide real-time pollution data and optimize control measures (Wang *et al.,* 2022).
- Expanding **international cooperation** through agreements like the Paris Agreement is crucial for enforcing stricter emission regulations and promoting the transfer of technology (Schnell & Prather, 2017).
- Encouraging **research and innovation in green technologies** such as carbon capture, advanced filtration, and sustainable energy solutions will drive progress in pollution control (Waseem *et al.,* 2023). Governments, industries, and researchers must collaborate to develop cost-effective solutions and ensure their accessibility worldwide. Addressing these challenges is critical to achieving long-term sustainability and improving global air quality.

CONCLUSION

Pollution in the atmosphere is indeed one of the primary global problems that affects human health and the environment. One of the best ways to overcome this challenge is by combining technology with policy enforcement and involving people's participation. Major reductions in emissions have been achieved through highly effective air pollution control technologies, such as filtration systems, electrostatic precipitators, scrubbers, catalytic converters, and other emerging green innovations. However, large-scale implementation of many of these technologies is still hindered by factors such as high costs, bad infrastructure, and political impediments. Aligning with standards and international cooperation, investments in the development of green technology will be crucial for the sustainable management of air pollution. Therefore, united efforts from around the globe have to be made to ensure clean air alongside a healthy future.

REFERENCES

Adepoju, A. O., Omotoso, I. O., Tiamiyu, O. G. (2023).

Bonjour, S., Adair-Rohani, H., Wolf, J., Bruce, N.G., Mehta, S., Prüss-Ustün, A., Lahiff, M., Rehfuess, E.A., Mishra, V., Smith, K.R. (2013). Solid fuel use for household cooking: country and regional estimates for

1980-2010. *Environ. Health Perspect., 121*(7), 784-790.
[http://dx.doi.org/10.1289/ehp.1205987] [PMID: 23674502]

Bui, M., Adjiman, C.S., Bardow, A., Anthony, E.J., Boston, A., Brown, S. (2018). Carbon capture and storage (CCS): the way forward. *Energy Environ. Sci., 11*(5), 1062-1176.
[http://dx.doi.org/10.1039/c7ee02342a]

Burnett, R., Chen, H., Szyszkowicz, M., Fann, N., Hubbell, B., Pope, C.A., III, Apte, J.S., Brauer, M., Cohen, A., Weichenthal, S., Coggins, J., Di, Q., Brunekreef, B., Frostad, J., Lim, S.S., Kan, H., Walker, K.D., Thurston, G.D., Hayes, R.B., Lim, C.C., Turner, M.C., Jerrett, M., Krewski, D., Gapstur, S.M., Diver, W.R., Ostro, B., Goldberg, D., Crouse, D.L., Martin, R.V., Peters, P., Pinault, L., Tjepkema, M., van Donkelaar, A., Villeneuve, P.J., Miller, A.B., Yin, P., Zhou, M., Wang, L., Janssen, N.A.H., Marra, M., Atkinson, R.W., Tsang, H., Quoc Thach, T., Cannon, J.B., Allen, R.T., Hart, J.E., Laden, F., Cesaroni, G., Forastiere, F., Weinmayr, G., Jaensch, A., Nagel, G., Concin, H., Spadaro, J.V. (2018). Global estimates of mortality associated with long-term exposure to outdoor fine particulate matter. *Proc. Natl. Acad. Sci. USA, 115*(38), 9592-9597.
[http://dx.doi.org/10.1073/pnas.1803222115] [PMID: 30181279]

Crutzen, P. J., Andreae, M. (1990). Biomass Buriing in the Tropics. *Impact on Atmospheric Chemistry and Biogeochemical Cycles, 250*(4988), 1669-1678.
[http://dx.doi.org/10.1126/science.250.4988.1669] [PMID: 17734705]

De Marco, A., Proietti, C., Anav, A., Ciancarella, L., D'Elia, I., Fares, S., Fornasier, M.F., Fusaro, L., Gualtieri, M., Manes, F., Marchetto, A., Mircea, M., Paoletti, E., Piersanti, A., Rogora, M., Salvati, L., Salvatori, E., Screpanti, A., Vialetto, G., Vitale, M., Leonardi, C. (2019). Impacts of air pollution on human and ecosystem health, and implications for the National Emission Ceilings Directive: Insights from Italy. *Environment, 125*, 320-333.
[http://dx.doi.org/10.1016/j.envint.2019.01.064] [PMID: 30739052]

Delhoménie, M.C., Heitz, M. (2005). Biofiltration of air: a review. *Crit. Rev. Biotechnol., 25*(1-2), 53-72.
[http://dx.doi.org/10.1080/07388550590935814] [PMID: 15999852]

Fayyazbakhsh, A., Bell, M.L., Zhu, X., Mei, X., Koutný, M., Hajinajaf, N., Zhang, Y. (2022). Engine emissions with air pollutants and greenhouse gases and their control technologies. *J. Clean. Prod., 376*, 134260.
[http://dx.doi.org/10.1016/j.jclepro.2022.134260]

Fenger, J. (1999). Urban air quality. *Atmos. Environ., 33*(29), 4877-4900.
[http://dx.doi.org/10.1016/S1352-2310(99)00290-3]

Franklin, B.A., Brook, R., Arden Pope, C., III (2015). Air pollution and cardiovascular disease. *Curr. Probl. Cardiol., 40*(5), 207-238.
[http://dx.doi.org/10.1016/j.cpcardiol.2015.01.003] [PMID: 25882781]

Ghulomjonovna, Y. Y. (2025). Engineering measures for the protection of atmospheric air: Methods for the purification of toxic gases. *American Journal Of Education And Learning, 3*(5), 232-237.
[http://dx.doi.org/10.5281/zenodo.15392748]

Goshua, A., Akdis, C.A., Nadeau, K.C. (2022). World Health Organization global air quality guideline recommendations: Executive summary. *Allergy, 77*(7), 1955-1960.
[http://dx.doi.org/10.1111/all.15224] [PMID: 35060140]

Hansel, N.N., McCormack, M.C., Kim, V. (2016). The Effects of Air Pollution and Temperature on COPD. *COPD, 13*(3), 372-379.
[http://dx.doi.org/10.3109/15412555.2015.1089846] [PMID: 26683097]

Hesterberg, T.W., Bunn, W.B., McClellan, R.O., Hamade, A.K., Long, C.M., Valberg, P.A. (2009). Critical review of the human data on short-term nitrogen dioxide (NO_2) exposures: evidence for NO_2 no-effect levels. *Crit. Rev. Toxicol., 39*(9), 743-781.
[http://dx.doi.org/10.3109/10408440903294945] [PMID: 19852560]

Kuklinska, K., Wolska, L., Namiesnik, J. (2014). Air quality policy in the U.S. And the EU – a review.

Atmos. Pollut. Res., 6(1), 129-137.
[http://dx.doi.org/10.5094/APR.2015.015]

Intergovernmental Panel on Climate Change. (2021). *Climate change 2021: The physical science basis: Contribution of Working Group I to the Sixth Assessment Report of the Intergovernmental Panel on Climate Change.* Cambridge University Press. https://www.ipcc.ch/report/ar6/wg1/

Masson-Delmotte, V., Zhai, P., Pirani, A., Connors, S. L., Péan, C., Berger, S., Caud, N., Chen, Y., Goldfarb, L., Gomis, M. I., Huang, M., Leitzell, K., Lonnoy, E., Matthews, J. B. R., Maycock, T. K., Waterfield, T., Yelekçi, O., Yu, R., & Zhou, B. (Eds.). (2021). *Climate change 2021: The physical science basis.* Cambridge University Press. https://www.ipcc.ch/report/ar6/wg1/

Jacobson, M.Z. (2017). Roadmaps to Transition Countries to 100% Clean, Renewable Energy for All Purposes to Curtail Global Warming, Air Pollution, and Energy Risk. *Earths Futur., 5*(10), 948-952.
[http://dx.doi.org/10.1002/2017EF000672]

Krigmont, H.V. (2023). Development of the Disinfecting Air Filter.*Proceedings of the 16th International Conference on Electrostatic Precipitation. ICESP 2022. Lecture Notes in Electrical Engineering.* (Vol. Vol. 1052). Cham: Springer.

Lelieveld, J., Evans, J.S., Fnais, M., Giannadaki, D., Pozzer, A. (2015). The contribution of outdoor air pollution sources to premature mortality on a global scale. *Nature, 525*(7569), 367-371.
[http://dx.doi.org/10.1038/nature15371] [PMID: 26381985]

Li, K., Wang, H., Li, J. (2022). Design and mechanism of photocatalytic oxidation for the removal of air pollutants: a review. *Environ. Chem. Lett., 20*, 2687-2708.
[http://dx.doi.org/10.1007/s10311-022-01436-7]

Liu, J., Tian, E., Zhang, S. (2023). The Rise of Two-Dimensional-Material-Based Filters for Airborne Particulate Matter Removal. *Adv. Fiber Mater., 5*, 461-483.
[http://dx.doi.org/10.1007/s42765-022-00242-8]

Liu, L., Cheng, S.Y., Li, J.B., Huang, Y.F. (2007). Mitigating Environmental Pollution and Impacts from Fossil Fuels: The Role of Alternative Fuels. *Energy Sources A Recovery Util. Environ. Effects, 29*(12), 1069-1080.
[http://dx.doi.org/10.1080/15567030601003627]

Matyssek, R., Schaub, M., Wieser, G. (2010). Air pollution and climate change effects on forest ecosystems: new evidence. *Eur. J. For. Res., 129*, 417-419.
[http://dx.doi.org/10.1007/s10342-010-0367-2]

Mills, G., Sharps, K., Simpson, D., Pleijel, H., Broberg, M., Uddling, J., Jaramillo, F., Davies, W.J., Dentener, F., Van den Berg, M., Agrawal, M., Agrawal, S.B., Ainsworth, E.A., Büker, P., Emberson, L., Feng, Z., Harmens, H., Hayes, F., Kobayashi, K., Paoletti, E., Van Dingenen, R. (2018). Ozone pollution will compromise efforts to increase global wheat production. *Glob. Change Biol., 24*(8), 3560-3574.
[http://dx.doi.org/10.1111/gcb.14157] [PMID: 29604158]

Molchanov, O., Krpec, K., Horák, J., Ochodek, T., Dej, M., Kubonová, L., Hopan, F., Ryšavý, J. (2023). Specifics of Electrostatic Precipitation of Fly Ash from Small- Scale Fossil Fuel Combustion. *Processes (Basel), 11*(3), 808.
[http://dx.doi.org/10.3390/pr11030808]

Montzka, S.A., Dlugokencky, E.J., Butler, J.H. (2011). Non-CO2 greenhouse gases and climate change. *Nature, 476*(7358), 43-50.
[http://dx.doi.org/10.1038/nature10322] [PMID: 21814274]

Neas, L.M. (2000). Fine particulate matter and cardiovascular disease. *Fuel Process. Technol., 65-66*, 55-67.
[http://dx.doi.org/10.1016/S0378-3820(99)00076-4]

Perera, F. P. (2017). Multiple threats to child health from fossil fuel combustion: Impacts of air pollution and climate change. *Environmental Health Perspectives* (2)
[http://dx.doi.org/10.1289/EHP299]

Perini, K., Magliocco, A. (2013). Effects of vegetation, urban density, building height, and atmospheric conditions on local temperatures and thermal comfort. *Urban For. Urban Green., 13*(3), 495-506. [http://dx.doi.org/10.1016/j.ufug.2014.03.003]

Schnell, J.L., Prather, M.J. (2017). Co-occurrence of extremes in surface ozone, particulate matter, and temperature over eastern North America. *Proc. Natl. Acad. Sci. USA, 114*(11), 2854-2859. [http://dx.doi.org/10.1073/pnas.1614453114] [PMID: 28242682]

Shindell, D., Lee, Y., Faluvegi, G. (2016). Climate and health impacts of US emissions reductions consistent with 2 °C. *Nature Climate Change, 6*, 503-507. Available form: https://doi.org/10.1038/nclimate2935 [http://dx.doi.org/10.1038/nclimate2935]

Suh, H. H., Bahadori, T., Vallarino, J., Spengler, J. D. (2000). Criteria air pollutants and toxic air pollutants. *Environmental Health Perspectives, 108*(Suppl. 4), 625-633. [http://dx.doi.org/10.1289/ehp.00108s4625]

Thompson, O.P., Breitenstein, R.S., Barnes, B.R., Glory, R. (2024). Promoting Behavioral Change and Public Awareness for Sustainable Clean Air.*Sustainable Strategies for Air Pollution Mitigation. The Handbook of Environmental Chemistry.* (Vol. Vol. 133). Cham: Springer.

Vig, N., Ravindra, K., Mor, S. (2023). Environmental impacts of Indian coal thermal power plants and associated human health risk to the nearby residential communities: A potential review. *Chemosphere, 341*, 140103.
[http://dx.doi.org/10.1016/j.chemosphere.2023.140103] [PMID: 37689154]

Wang, L., Xie, Q., Xue, F., Li, Z. (2022). Does Smart City Construction Reduce Haze Pollution? *Int. J. Environ. Res. Public Health, 19*(24), 16421. [http://dx.doi.org/10.3390/ijerph192416421] [PMID: 36554300]

Waseem, M., Amir, M., Lakshmi, G.S., Harivardhagini, S., Ahmad, M. (2023). Fuel cell- based hybrid electric vehicles: An integrated review of current status, key challenges, recommended policies, and future prospects. *Green Energy and Intelligent Transportation, 2*(6), 100121. [http://dx.doi.org/10.1016/j.geits.2023.100121]

Wyer, K.E., Kelleghan, D.B., Blanes-Vidal, V., Schauberger, G., Curran, T.P. (2022). Ammonia emissions from agriculture and their contribution to fine particulate matter: A review of implications for human health. *J. Environ. Manage., 323*, 116285. [http://dx.doi.org/10.1016/j.jenvman.2022.116285] [PMID: 36261990]

Ye, Z., Yang, J., Zhong, N., Tu, X., Jia, J., Wang, J. (2020). Tackling environmental challenges in pollution controls using artificial intelligence: A review. *Sci. Total Environ., 699*, 134279. [http://dx.doi.org/10.1016/j.scitotenv.2019.134279] [PMID: 33736193]

CHAPTER 7

The Association between Carbon Emissions and the Human Capital Index in Emerging Nations

Md. Atikur Rahman[1], **Md. Afzal Hossain**[2,*] and **Sohel Rana**[2]

[1] *School of Management, Jiujiang University, Jiujiang , China*

[2] *College of Business Administration, International University of Business Agriculture and Technology, Dhaka, Bangladesh*

Abstract: This study examined the empirical cointegration, long-term, and short-term relationships, as well as the causality nexus, between the Human Development Index and CO_2 emissions in Bangladesh from 1990 to 2019. It used the ARDL Bounds test and the Granger causality framework. The ARDL F-statistic bounds tests disclose long-run confirmation of cointegration. The human development index has time-invariant impacts on the growth of CO_2 emissions; *i.e.*, these effects are consistent in both the long and short term. However, there is no proof of a causal relationship between HDI and CO_2 emissions. The robustness of the results was tested using two alternative assessment methods: fully modified least squares and dynamic least squares. The Human Development Index is statistically significant. This paper provides recommendations for the government on CO_2 emission reduction measures and serves as a valuable decision-making tool for accelerating low-carbon development.

Keywords: CO_2 emissions, Energy Consumption, Granger causality test, Human Development Index, The ARDL bounds tests.

INTRODUCTION

Although developing nations are encouraged to join efforts to reduce CO_2 emissions and avert hazardous climate change, the impact of CO_2 reduction targets on the human development of developing nations remains a topic of debate. For a consensus to emerge in favor of low CO_2 stabilization objectives, developing nations must participate in global emission reduction efforts. To keep global temperatures from rising by more than 2°C, it has been argued, for example, that developing countries should achieve CO_2 reductions of more than 20% below business-as-usual levels. (Hossain and Chen, 2020; Banu *et al.*, 2018).

[*] **Corresponding author Md. Afzal Hossain**: College of Business Administration, International University of Business Agriculture and Technology, Dhaka, Bangladesh; E-mail: afzal.cba@iubat.edu

Sanjeev Kumar, Mohammad Badruddoza Talukder & A.K. Haghi (Eds.)

Because developing nations are expected to rely heavily on fossil energy to meet their development needs, the potential consequences of such decreases on development criteria remain unknown. Developing countries have expressed concerns about whether development goals can be achieved in light of current technological advancements and population patterns (Hossain *et al.,* 2021b; Khan *et al.,* 2021; Wang *et al.,* 2018).

In several ways, this study advances our current understanding of the field.

As it is clear that all of these variables are interconnected, this research is, first and foremost, a first step in the ongoing effort to quantify the relationship between HDI and CO_2 emissions. Further research is needed in this area. However, it could be important to quantify the link between CO_2 emissions and HDI. The investigation of the relationship between HDI and CO_2 emissions is a novel and innovative piece of study in the area.

Second, when it comes to characterizing the human development process, this research is the first to apply the ARDL model and Granger causality in the context of a Bangladeshi setting. Bangladesh's current situation in terms of energy and human development is fragile, and the nation's energy issues are damaging both the economy and humanity as a whole. Bangladesh boasts a wealth of natural resources, and investors can capitalize on the use of renewable energy sources.

Third, policymakers must consider the environmental, economic, and social aspects of sustainable development. However, past studies have concentrated on how energy usage affects economic development rather than how it affects people's well-being. Investigating the link between the studied variables would provide policymakers with more data and a more comprehensive understanding.

Objective of the Study

1. To Examine the Relationship Between CO_2 Emissions and Human Development Index (HDI) in Bangladesh
2. To Explore the Role of Energy Consumption in Human Development and Environmental Degradation
3. To Provide Policy Recommendations for Sustainable Development
4. To Contribute to the Literature on CO_2 Emissions and HDI
5. To Evaluate Bangladesh's Progress in Human Development and Environmental Sustainability

Theoretical Overview

1. **Human Development Index (HDI) as a Measure of Development:** The HDI, introduced by the UNDP in 1990, serves as a comprehensive indicator of

human development, encompassing three key dimensions: access to knowledge (education), long and healthy lives (health), and decent living standards (income). Unlike GDP, which focuses solely on economic output, the HDI incorporates social and health metrics, providing a more holistic view of development. Recent adaptations of the HDI, such as the Planetary-adjusted HDI (PHDI), also consider environmental factors, including CO_2 emissions, reflecting the increasing importance of sustainability in development frameworks.

2. **CO_2 Emissions and Development:** CO_2 emissions are a critical driver of climate change, with developing nations like Bangladesh facing the dual challenge of reducing emissions while pursuing economic growth. Fossil fuel consumption remains a primary source of CO_2 emissions, and its reduction is essential to limit global temperature rise to below 2°C. However, developing nations often rely on fossil fuels to meet their energy needs, raising concerns about the trade-offs between emission reduction targets and human development goals.
3. **Energy Consumption and Human Development:** Energy consumption is a fundamental driver of both economic growth and human development. However, excessive reliance on non-renewable energy sources exacerbates environmental degradation, particularly through increased CO_2 emissions. Balancing energy consumption with sustainable practices is crucial for achieving long-term human development without compromising environmental health.
4. **Sustainable Development and Policy Implications:** Sustainable development requires integrating environmental, economic, and social dimensions. Policymakers must prioritize strategies that promote renewable energy, reduce CO_2 emissions, and enhance human welfare. The case of Bangladesh highlights the urgency of addressing energy consumption patterns and their impact on both human development and environmental sustainability.
5. **ARDL Framework and Granger Causality:** The ARDL (Autoregressive Distributed Lag) model is a robust statistical tool for analyzing long-term and short-term relationships between variables, particularly in time-series data. Granger causality analysis complements this by identifying causal relationships between variables, such as CO_2 emissions and HDI. These methodologies provide a rigorous basis for understanding the interplay between environmental and developmental factors.
6. **Bangladesh as a Case Study:** Significant increases in CO_2 emissions and energy consumption have accompanied Bangladesh's rapid economic growth. Despite improvements in HDI, the country faces challenges in reconciling development goals with environmental sustainability. This study highlights the

need for targeted policies to address these challenges and offers a model for other developing nations facing similar issues.

Literature Review

The primary goal of economic growth is human development, measured by the Human Development Index (HDI). To achieve a higher HDI, it is essential to identify the factors that influence human development. Recent research has explored various determinants, including environmental degradation, energy consumption, and economic growth, using diverse samples, econometric models, and explanatory variables.

A significant body of research focuses on the Environmental Kuznets Curve (EKC) hypothesis, which posits an inverted U-shaped relationship between environmental degradation and per capita income. Initially, pollution increases with rising income, but after a certain threshold, environmental degradation begins to decrease as income continues to grow. Studies by Alkhathlan and Javid (2013) in Saudi Arabia, Saboori and Sulaiman (2013) in Malaysia, Cil Yavuz (2014) in Turkey, Ozcan (2013) in Middle Eastern countries, and Soytas *et al.*, (2007) in the United States have examined the validity of the EKC hypothesis. However, this study diverges from the EKC framework by focusing on causality and empirical evidence, aligning with the view that econometric analysis should not solely rely on existing theories but also contribute to theory-building, especially in the context of environmental degradation and inclusive development in developing nations.

Environmental degradation directly impacts HDI, which comprises income, health, education, and life expectancy. Energy consumption is another critical factor influencing human development. While increased energy use was traditionally associated with improved human development, growing environmental concerns have challenged this perspective. Tran *et al.*, (2019) argue that higher energy consumption does not necessarily lead to better human development, particularly in the context of ecological sustainability.

Several studies have investigated the relationship between energy consumption and HDI. Martinez and Ebenhack (2008) found a strong correlation between energy use and HDI across 120 countries. Ouedraogo (2013) examined 15 developing countries from 1988 to 2008, identifying unidirectional Granger causality from energy consumption to human development. In contrast, Tran *et al.*, (2019) analyzed 93 countries from 1990 to 2014 and found no positive correlation between energy consumption and human development in both developed and developing economies. Similarly, Niu *et al.*, (2013) used Granger panel causality tests for 50 countries from 1990 to 2009, revealing bidirectional causality between electricity consumption and HDI. Conversely, Khan *et al.*, (2018) found no significant impact of electricity consumption on HDI in Pakistan from 1990 to 2016 using ARDL bounds analysis. Wang *et al.*, (2018) observed a decline in Pakistan's HDI from 1990 to 2014, which they attributed to the adoption of renewable energy. In contrast, Shahbaz and Lean (2012) identified a feedback effect between energy consumption and economic growth in Pakistan.

Research on the nexus between CO_2 emissions and HDI is limited but significant. Alam *et al.*, (2012) studied Bangladesh, finding long-term unidirectional causality from economic growth to energy consumption and bidirectional causality between economic growth and electricity consumption, with no short-term causal relationships. Analyzed Bangladesh from 1990 to 2018, identifying weak and strong decoupling between CO_2 emissions and HDI. Their index decomposition revealed that economic activity was the primary driver of CO_2 emissions.

In summary, the relationship between energy consumption, environmental degradation, and human development is complex and context-dependent. While energy consumption is crucial for economic growth, its impact on HDI varies across countries and time periods. Environmental degradation, particularly CO_2 emissions, poses significant challenges to human development, especially in developing nations like Bangladesh. Policymakers must strike a balance between economic growth, energy consumption, and environmental sustainability to achieve long-term human development goals. Empirical studies, such as those using ARDL bounds analysis and Granger causality tests, provide valuable insights but also highlight the need for further research to address gaps and inconsistencies in the literature.

Methodology

Variables Explanation

Four variables were employed in the growth mode to make operational assessments of long-term and short-term dynamic forces, as well as the relationship between energy consumption, HDI, GDP, and CO_2 emissions. In growth mode, applying a variable provides data concerning the variable's trend of actions in the current year compared to the previous year, exposing dynamic relationships between related variables that can be used to get useful insight into the variable's future trend. In this study, the rise of CO_2 emissions is regarded as a dependent variable. Whereas HDI, GDP, and energy consumption are explanatory variables. The study is based on 30 years of data (1990-2019). Table **1** lists all variables and their descriptions.

Table 1. Variable description.

Variable	Measurement	Data Source
ENRE	Energy consumption: kilograms equivalent to oil	International Energy Agency
GDP	Gross domestic product: (2010 constant US$ $)	World Development Indicators(WDI)
HDI	Human Development Index	UNDP
CO_2	CO_2 dioxide emissions: Kilo Tonne	WDI

Source: Author's compilation based on Soytas *et al.*, (2007)

The model's operating formula is listed below.

CO_2 Emissions f (HDI, ENRE, GDP)

The model's econometric formula is given below (Eq. **1**).

$$CO_{2t} = \alpha + \beta_1 HDI_t + \beta_2 ENRC_t + \beta_3\, GDP + \varepsilon_t \quad \textbf{(1)}$$

α is the constant term, β_1, β_2, β_3 are the slope parameters, and ε is the error expression.

The Unit Root Check

The ARDL cointegration approach does not require pre-testing for unit roots, allowing variables of order I(0) or I(1). However, variables must not be I(2), as this invalidates the method. Stationarity was assessed using Augmented Dickey-Fuller (ADF) and Phillips-Perron (PP) tests before proceeding with the ARDL Bounds analysis (Eq. **2**). The ARDL technique for cointegration is as follows for testing cointegration bounds

$$\Delta CO_{2t} = \alpha + \sum_{i=0}^{p} \beta_i \Delta CO_{2t-i} + \sum_{i=0}^{q} \partial_i \Delta HDI_{t-i} + \sum_{i=0}^{r} \gamma_i\, \Delta ENRC_{t-i} + \sum_{i=0}^{s} \delta_i \Delta GDP_{t-i} + \theta_0 CO_{2t-1} + \theta_1\, HDI_{t-1} + \theta_2 ENRC_{t-1} + \theta_3 GDP_{t-1} + \varepsilon_{t1} \ldots\ldots \quad \textbf{(2)}$$

Where CO_2, HDI, ENRC, and GDP are study variables, and εt_1 is a random "disturbance" expression.

(Eq. **2**) 's Error Correction Method (ECM) is a type of ECM in which the coefficients are not limited. Pesaran *et al.*, (2001) refer to it as a "conditional ECM."The four terms with summation signs in Eq. (**2**) indicate the dynamics of mistake correction (2). In contrast, the second section (terms beginning with a θ) refers to the long-term relationship (Shahbaz *et al.,* 2013). The suitable standards to regulate lags, p, q, r, and s would be calculated using one or more of the "evidence benchmarks" – HQ, BIC, AIC, and so on.

Based on the aforementioned equation, the null and alternative hypotheses are provided below:

H0: Cointegration does not exist.

H1: Cointegration exists.

The null hypothesis is verified using an F-test for the cumulative consequence of the coefficients of the lagged levels of the variables. Therefore

$$H_0: \theta_1 = \theta_2 = \theta_3 = 0$$

$$H_1: \theta_0 \neq 0, \theta_1 \neq 0, \theta_2 \neq 0, \theta_3 \neq 0$$

The ARDL bounds test uses a non-standard F-statistic distribution, with critical values established by Pesaran *et al.*, (2001) for I(0) and I(1) variables. For smaller samples, Narayan (2005)'s critical values are more appropriate. If the F-statistic is below the lower bound, variables are I(0), indicating no cointegration. If it exceeds the upper bound, cointegration exists. If it falls within the bounds, the test is inconclusive and requires additional cointegration methods. A "Bounds t-test" is also recommended for verification, as noted by Giles (2013).

$$H_0: \theta_0 = 0, \quad against\ H_1: \theta_0 < 0$$

This test's decision rule is like this:

In Eq. (**1**) for CO_{2t}-1, if the t-statistic is larger than Pesaran *et al.*, (2001)'s "I(1) bound," the conclusion that the variables have a long-term connection is supported. If the "I(0) bound" is less than the t-statistic, this study may assume that all of the data is stationary. The systematic ECM is used to estimate short-run parameters, as shown in Eq. (**3**) below:

$$\Delta CO_{2t} = \alpha + \sum_{i=1}^{p} \beta_i \Delta CO_{2t-i} + \sum_{i=0}^{q} \partial_i \Delta HDI_{t-i} + \sum_{i=0}^{r} \gamma_i \Delta ENRC_{t-i} + \sum_{i=0}^{s} \delta_i \Delta GDP_{t-i} + +\tau ECT_{t-1} + \varepsilon_{t3} \quad \textbf{(3)}$$

The outcomes of the ECM indicate how quickly short-term shocks return to long-run symmetry. The ECM combines short-term and long-term coefficients without losing long-term data. Long-term causality is revealed by the substantial value of the error correction term (ECT) coefficient in the ECM technique, whereas short-term causality is revealed by the major value of additional descriptive factor coefficients (Shahbaz *et al.,* 2013; Rahman & Mamun, 2016**).**

The Diagnostic Tests of the Model

The ARDL Bounds analysis assumes the error term is normally distributed and serially independent. Serial independence is tested using the Breusch-Godfrey LM test, model specification with the Ramsey RESET test, and heteroscedasticity with the Breusch-Pagan-Godfrey test to ensure the validity of the regression model.

The Stability Test of the Model

The dynamic stability of any model having an autoregressive form must be verified. The stability of the model will be evaluated using the CUSUM and CUSUM of squares tests (Brown *et al.,* 1975). Pesaran & Pesaran (1997) also offer these studies to assess constraint consistency.

Robustness Check

The FMOLS and DOLS procedures are used to evaluate the validity of the results.

Granger Causality Test

If time series are cointegrated, Granger causality exists in at least one direction. Correlation alone is insufficient, as it may be spurious or influenced by unobserved variables, and does not imply causality. For cointegrated series, causality must be verified using a VAR model to ensure accurate conclusions about directional relationships (Eqs.4 and 5)..

$$Y_t = g_0 + a_1 Y_{t-1} + \ldots + a_p Y_{t-p} + b_1 X_{t-1} + \ldots + b_p X_{t-p} + \mu_t \quad (4)$$

$$X_t = h_0 + c_1 X_{t-1} + \ldots + c_p X_{t-p} + d_1 Y_{t-1} + \ldots + d_p Y_{t-p} + \vartheta_t \quad (5)$$

To test H_0: $b_1 = b_2 = \ldots = b_p = 0$, compared to H_A: Granger X is the cause of Y. Likewise, to test H_0: $d_1 = d_2 = \ldots = d_p = 0$, compared to H_A: Granger, Y is the cause of X. When the null hypothesis is rejected, Granger causality is indicated in each case. The X and Y sequences are presented in a 'level' format, which essentially indicates that the figures are not in a 'difference' format. White noise error expressions are represented by by μt and ϑt.

These errors should be zero in the long-run equilibrium. If statistically no less than one of the d_i or b_i coefficients is distinct from 0 in the two equations, the sequences X_t and Y_t are cointegrated. In the long run, if di = 0 and bi ≠ 0, X_t will outstrip Y_t. If di ≠ zero and bi = 0, the opposite will happen. When di‡0 and bi‡0 are together, subsequently, X_t and Y_t have a feedback connection. However, there is no cointegration between X_t and Y_t if di and bi are both 0. When the sample size is insufficient to satisfy the asymptotic requirements for cointegration and causality assessments, contradicting conclusions (as seen in the prior result of ARDL) can emerge (Giles, 2011). The dynamics of Y_t and X_t in the short term are represented by the coefficients ai's and ci's. In the short term, X_t movement will lead to Y_t movement if the ai's are not entirely zero, and in the long term, Y_t motions will result in X_t if the ci's are not entirely zero. Using a Vector Autoregression (VAR) structure, the Toda-Yamamoto (1995) method will be used to compute the Granger Causality between the variables. Using standard methods, this study will determine the suitable maximum lag limit for the variables in the VAR. Table **2** discusses the duration of the lag, which is determined using the accepted information criterion, such as AIC. We will also make sure VAR is well-specified, meaning the residuals do not contain serial correlation.

FINDINGS AND DISCUSSION

The Descriptive Indicator Statistics

Table **3** provides a summary of the descriptive indicator statistics for all the variables used to achieve our goal. Equally, it is established that the maximum point of the standard deviation is from CO_2 emissions, followed by energy consumption, GDP, and HDI, respectively. Table **3** mentions that the skewness measures indicate that all variables in this study are properly symmetric, except for energy use, which is strongly skewed to the right. Furthermore, as shown by the Kurtosis measurement, platykurtic distributions were employed for each variable in this investigation. The Jarque-Bera numbers and the p-values are in favor of rejecting the null hypothesis that a standard distribution exists.

Table 2. The descriptive indicator statistics.

-	HDI	CO_2	GDP	ENERGY
Mean	-0.29	-0.53	0.74	2.28
Median	-0.29	-0.55	0.74	2.22
Maximum	-0.20	-0.07	0.90	2.65
Minimum	-0.40	-0.86	0.54	2.07
Std. Dev.	0.06	0.22	0.09	0.17

(Table 2) cont.....

-	HDI	CO_2	GDP	ENERGY
Skewness	-0.14	0.23	-0.25	0.88
Kurtosis	1.90	2.06	2.47	2.59
Jarque-Bera	1.61	1.38	0.67	4.11
Probability	0.44	0.50	0.71	0.12
Sum	-8.85	-15.98	22.32	68.43
Sum Sq. Dev.	0.11	1.45	0.24	0.83
Observations	30	30	30	30

Table 3. The outcome of the unit root analysis Panel A: The unit root test (Phillips & Perron, 1988).

Variable	-	CO_2	ENERGY	GDP	HDI
With Constant	t-Statistic	1.89	2.29	-2.14	-1.68
-	*Prob.*	*0.99*	*0.99*	*0.22*	*0.42*
-	-	n0	n0	n0	n0
With Constant & Trend	t-Statistic	-0.90	-0.24	-5.44	-2.91
-	*Prob.*	*0.94*	*0.98*	*0.00*	*0.17*
-	-	n0	n0	***	n0
Without Constant & Trend	t-Statistic	-1.58	4.81	0.11	-9.10
-	*Prob.*	*0.10*	*1.00*	*0.71*	*0.00*
-	-	n0	n0	n0	***

The Outcome of Unit Root Analysis

Tables **4**, **5** and **6** examined the long-term nexus between energy consumption, CO_2 emissions, GDP, and HDI using the ARDL bound estimate approach. A bound estimate has the benefit of being flexible in terms of the series' order of integration. The variables must therefore be integrated into either I (0) or I because of this (1). Both the PP estimation (Phillips & Perron, 1988) and the ADF test (Dickey & Fuller, 1981) were used to determine the order for all variables and to assess the persistence of the unit root test. As indicated in Table **3**, each variable employs a unit root test in its levels, even though it is stable in its first differences. It implies that each series of order I (1) is integrated.

VAR Selection Criteria of Lag Limit

Before calculating the ARDL F bound assessment, the ARDL bounds check must be completed. The appropriate lag sequence of the variables must be determined to assess whether the variables are cointegrated, as discussed in Table **7**. When choosing a lag length order, the estimate of the F-statistic presents a serious issue.

The endogenous variables of CO_2 emissions, HDI, GDP, and energy consumption were included in the VAR technique, which was primarily used in the study to determine the optimal potential lag limit. A diagnostic assessment of the lag length period was then applied to the residuals. Table **4** displays the measurements from this calculation.

Table 4. Panel B: The unit root test (Phillips & Perron, 1988) first difference.

Variable	-	CO_2	ENERGY	GDP	HDI
With Constant	t-Statistic	-4.72	-3.53	-9.19	-5.16
-	*Prob.*	*0.00*	*0.01*	*0.00*	*0.00*
-	-	***	**	***	***
With Constant & Trend	t-Statistic	-5.20	-4.17	-8.94	-5.03
-	*Prob.*	*0.00*	*0.01*	*0.00*	*0.00*
-	-	***	**	***	***
Without Constant & Trend	t-Statistic	-0.61	-2.37	-9.04	-0.55
-	*Prob.*	*0.44*	*0.01*	*0.00*	*0.46*
-	-	n0	**	***	n0

Table 5. Panel C: The unit root test (Dickey & Fuller, 1981).

Variable	-	CO_2	ENERGY	GDP	HDI
With Constant	t-Statistic	3.01	3.94	-2.09	-2.87
-	*Prob.*	*1.00*	*1.00*	***0.24***	***0.06***
-	-	n0	n0	n0	*
With Constant & Trend	t-Statistic	-0.88	0.10	-5.56	-2.72
-	*Prob.*	*0.94*	*0.99*	***0.00***	***0.23***
-	-	n0	n0	***	n0
Without Constant & Trend	t-Statistic	-3.68	4.22	0.28	-15.99
-	*Prob.*	*0.00*	*1.00*	***0.76***	***0.00***
-	-	***	n0	n0	***

Table 6. Panel D: The unit root test (Dickey & Fuller, 1981) first difference.

Variable	-	CO_2	ENERGY	GDP	HDI
With Constant	t-Statistic	-4.67	-3.56	-19.27	-5.34
-	*Prob.*	*0.00*	*0.01*	***0.00***	***0.00***
-	-	***	**	***	***
With Constant & Trend	t-Statistic	-5.33	-4.56	-18.54	-7.48

(Table 6) cont.....

Variable	-	CO_2	ENERGY	GDP	HDI
-	*Prob.*	*0.00*	*0.00*	***0.00***	***0.00***
-	-	***	***	***	***
Without Constant & Trend	t-Statistic	-2.14	-2.37	-10.82	-2.17
-	*Prob.*	*0.03*	*0.01*	***0.00***	***0.03***
-	-	**	**	***	**

Table 7. Lag limit VAR selection criterion.

Lag	LogL	LR	FPE	AIC	SC	HQ
0	187.23	NA	2.43e-11	-13.08	-12.89	-13.02
1	302.21	188.90*	2.10e-14*	-20.15*	-19.20*	-19.86*
2	313.89	15.85	3.09e-14	-19.84	-18.13	-19.32

Result of Long-term Assessments of ARDL and Bounds Check

In Tables **8** and **9**, which discuss the results of the method's approval of all the analytical evaluations, we can now proceed to the next stage of the research, including border and cointegration checks. The long-run findings are presented in Table **5**, along with the results of the ARDL limits F-test cointegration method.CO_2 emissions have a one-lag length, whereas HDI, energy consumption, and GDP have a zero-lag length. The value of the F-statistic is 13.20, which is greater than the 1%, 5%, and 10% critical values for the 0.05 and 0.10 significance levels, respectively. The higher range of F-statistics validates the cointegration nexus between CO_2 emissions, energy consumption, HDI, and GDP in Bangladesh. Following that, F-statistic bounds tests confirm cointegration, revealing a long-term connection. The long-term impacts reveal significant correlations between energy usage, HDI, and CO_2 emissions. This suggests that long-term reductions in CO_2 emissions are a result of HDI and energy usage, as evidenced by the symbol and statistical implications of their coefficients.

Table 8. Long-term estimates of ARDL and bounds check.

Variable	Coefficient	Std. Error	t-Statistic	Prob.
ENERGY	0.64	0.28	2.22	0.03
GDP	0.08	0.26	0.34	0.73
HDI	2.18	0.71	3.04	0.00
C	-1.34	0.84	-1.58	0.12

Table 9. F-Bounds Check.

Test Statistic	Value	Signif.	I(0)	I(1)
F-statistic	13.20	10%	2.38	3.1
K	3	5%	2.78	3.68
-	2.5%	3.16	4.07	2.5%
-	1%	3.65	4.66	1%

The Short-run Analysis

Table **10** reveals that the projected Error Correction Model (ECM) has a negative coefficient, which is statistically significant. These calculations indicate that the long-term equilibrium point is achieved by adjusting any deviation from a long-term balance of the variables by approximately 53% for each period. In the ARDL setting, short-run causation is confirmed. The results are shown in Table **6**. In conjunction with the Lag periods of HDI and energy consumption, the symbol and statistical implications of the coefficients indicate a short-run effect on CO_2 emissions. Therefore, the overall influence of reciprocally developing HDI and energy consumption on the progression of CO_2 emissions is time-invariant, *i.e.*, they have equivalent long-term and short-term effects.

Table 10. The short-term study.

Variable	Coefficient	Std. Error	t-Statistic	Prob.
D(ENERGY)	0.34	0.16	2.14	0.04
D(GDP)	0.02	0.05	0.35	0.72
D(HDI)	1.88	0.58	3.23	0.00
ECT	-0.53	0.18	-2.91	0.00

The Analytical Test

Tables **11**, **12**, and **13** present the LM test, the Heteroskedasticity test for measurement, and the RESET assessment for the normality of the residual component. The statistics indicate that the short-term technique permits complete analytical calculations. Serial correlation, which utilizes normally distributed residual expressions, and heteroscedasticity, the method's operational formula, are not commonly illustrated in the literature. In Fig. (**1**), the p-values of the LM, heteroscedasticity, and Ramsey reset assessments are all greater than 0.1, indicating no effect on serial correlation, heteroscedasticity, or the model's operational formula. To ensure consistency between the long-term and short-term limits, the CUSUM and CUSUMSQ evaluations were employed (Pesaran *et al.*, 2001). The data is represented as a graph, and Figs. (**2** and **3**) show, respectively, the outlines for the CUSUM and CUSUMSQ tests. Fig. (**2**) explains the balance between the diagnostic boundaries at the 5% level; the cumulative total and cumulative sum of squares plots are also equal.

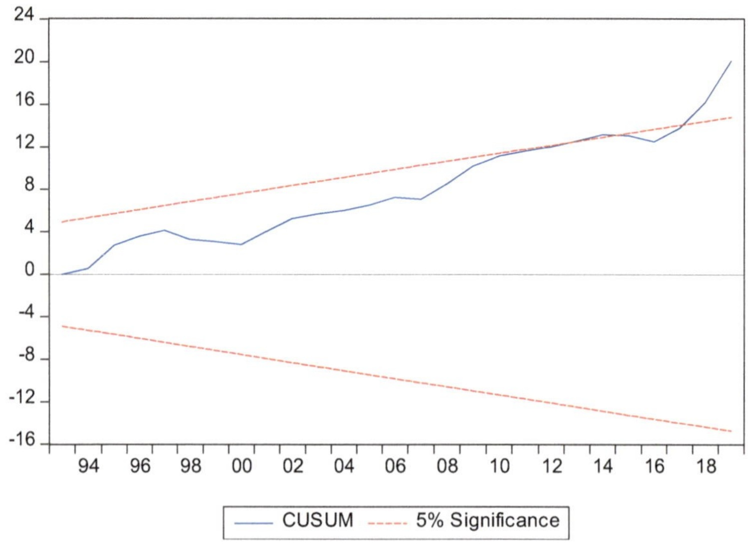

Fig. (1). Plots of the cumulative sum. **Source:** Author's creation.

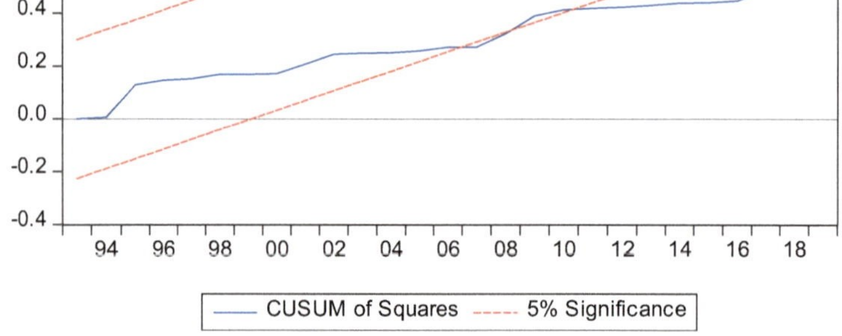

Fig. (2). Plots of the cumulative sum of squares. **Source:** Author's creation.

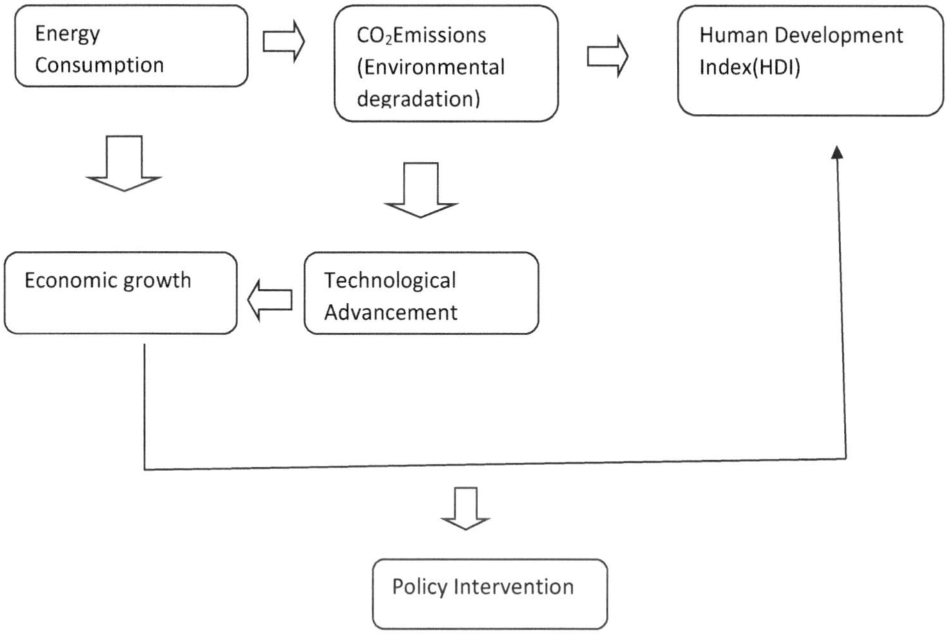

Fig. (3). Visual representation of the conceptual framework.
Source: Author's compilation based on Hossain *et al*., (2020)

Table 11. LM Test: Breusch-Godfrey.

LM Test: Breusch-Godfrey			
F-statistic	6.753896	Prob. F(2,25)	0.4511
Obs*R-squared	10.52342	Prob. Chi-Square(2)	0.2371

Table 12. Heteroskedasticity Test: Breusch-Pagan-Godfrey.

Heteroskedasticity Test: Breusch-Pagan-Godfrey			
F-statistic	4.954686	Prob. F(3,26)	0.1775
Obs*R-squared	10.91232	Prob. Chi-Square(3)	0.1122
Scaled explained SS	14.22569	Prob. Chi-Square(3)	0.0026

Table 13. The analytical test.

RESET Test: Ramsey			
-	Value	df	Probability
t-statistic	4.013605	26	0.3715
F-statistic	16.10902	(1, 26)	0.3715
Likelihood ratio	14.46497	1	0.3669

The Result of the Granger Causality Checks

TabTableus, causality test, which was used to analyze the causal relationships between CO_2 emissions, energy use, GDP, and HDI in Bangladesh. Results revealed unidirectional causality from GDP and energy use to CO_2 emissions, indicating that economic growth and energy consumption are driving factors in CO_2 emissions. However, no causality was found from HDI to CO_2 emissions. These findings align with studies by Lean and Smyth (2009), Shahbaz *et al.*, (2016), and Mahmud and Shahab (2014), but contradict those of Farhani and Rejeb (2012). The study concludes that rising energy use and GDP are key drivers of increasing CO_2 emissions in Bangladesh, with no significant influence from HDI as shown in the Table **14**.

Table 14. Source: Authors compiled from Shahbaz *et al.* (2016).

Null Hypothesis:	Obs	F-Statistic	Prob.
Energy consumption does not Granger-cause CO_2 emissions	28	3.54	0.04
CO_2 emissions do not Granger-cause energy consumption	28	1.10	0.34
GDP does not Granger-cause CO_2 emissions	28	0.49	0.61
CO_2 emissions do not Granger-cause GDP	28	4.06	0.03
HDI does not Granger-cause CO_2 emissions	28	0.18	0.83
CO_2 emissions do not Granger-cause HDI	28	1.01	0.37

Robustness Check

It examines the reliability of the results; FMOLS and DOLS, two alternative assessment procedures, are applied. Tables **9** and **10** present the experimental findings, including the impact of energy use and HDI on CO_2 emissions. Energy use and HDI are both statistically significant, as mentioned in Fig. (**3**). As the FMOLS and DOLS models are considered to be robust in their logic of incorporating both lags and leads, as well as accounting for autocorrelations and heteroscedasticity, the findings produced by these techniques are similar to those obtained from the ARDL model As mentioned in the Table **15** and **16** (Salim, Yao, & Chen, 2017).

Table 15. Source: Authors compiled based on Salim, Yao, & Chen (2017).

Variable	Coefficient	Std. Error	t-Statistic	Prob.
ENERGY	0.41	0.08	4.95	0.00
GDP	0.08	0.09	0.86	0.39
HDI	2.47	0.24	9.93	0.00

(Table 15) cont.....

Variable	Coefficient	Std. Error	t-Statistic	Prob.
C	-0.81	0.26	-3.09	0.00

Table 16. Source: Authors compiled based on Salim, Yao, & Chen (2017).

Variable	Coefficient	Std. Error	t-Statistic	Prob.
ENERGY	0.40	0.09	4.09	0.00
GDP	0.00	0.15	0.00	0.99
HDI	2.67	0.25	10.61	0.00
C	-0.62	0.27	-2.28	0.03

SOLUTIONS AND RECOMMENDATIONS

The following courses of action for reducing CO_2 emissions and achieving sustainable ecological progress in Bangladesh can be adopted based on the study's empirical findings.

1. In Bangladesh, CO_2 emissions are strongly linked to HDI levels. As a result, focusing solely on fuel and ecological techniques to reduce CO_2 emissions will not effectively advance the growth plan to reduce air pollution; its HDI level will remain at the bottom line.
2. Short-term plans to promote HDI should be developed, with a focus on satisfying the necessary prerequisites of economic growth, human health, and education before addressing CO_2 emissions.
3. The country's human growth is also hampered by trade. In Bangladesh, the majority of the workforce is employed in the agricultural industry, which has remained unaddressed by the government for the past decade.
4. The government should focus primarily on the country's agricultural industry. It is recommended that small-scale industrial ventures be established to help underprivileged individuals in rural areas earn a higher income.
5. In Bangladesh, poverty is the key factor limiting human development, and economic inequality is one of the main drivers of rising poverty. One potential issue that has to be addressed is how poverty and economic inequality affect the nation's process of human development. The rich get richer, and the poor get poorer.
6. To alleviate Bangladesh's energy shortages, the government must utilize renewable energy resources. The Bangladeshi government can achieve its Sustainable Development Goals (SDGs) and ensure environmental sustainability by addressing the issue of energy crises. The Bangladeshi government must combine these goals with the adoption of renewable energy sources to achieve this.

7. Longevity, education, and standard of living are three well-known criteria that are included in the HDI. Energy has an impact on all of these traits, which is one of the reasons why nations should concentrate on reducing greenhouse gas emissions and enhancing energy efficiency.
8. By requiring businesses to utilize cutting-edge equipment, eco-friendly production techniques, and labor training, while also raising awareness about the harmful effects of CO_2 emissions, the Bangladeshi government can enforce environmental regulations, increase production, accelerate economic development, and reduce CO_2 emissions.

CONCEPTUAL FRAMEWORK

1. **Energy Consumption → CO_2 Emissions:**
2. Increased energy consumption, particularly from fossil fuels, results in higher CO_2 emissions, which contribute to environmental degradation.
3. **CO_2 Emissions → HDI:**
4. High CO_2 emissions can negatively impact human development by exacerbating climate-related health risks, reducing agricultural productivity, and increasing vulnerability to extreme weather events.
5. **Energy Consumption → HDI:**
6. Energy consumption supports economic activities and infrastructure development, which are essential for improving health, education, and living standards. However, unsustainable energy use can undermine long-term human development.
7. **Economic Growth (GDP) → HDI:**
8. Economic growth provides resources for investments in health, education, and social services, directly enhancing HDI.
9. **Technological Advancements → CO_2 Emissions and HDI:**
10. Technological innovations in renewable energy and energy efficiency can reduce CO_2 emissions while supporting sustainable human development.
11. **Policy Interventions → CO_2 Emissions, Energy Consumption, and HDI:**
12. Effective policies can promote renewable energy, reduce emissions, and allocate resources to improve human development outcomes.

CONCLUSION

This study examined the relationship between HDI and CO_2 emissions in Bangladesh (1990–2019) using ARDL bounds analysis and Granger causality. The ARDL test confirmed cointegration, revealing that HDI and energy consumption have consistent long- and short-term effects on CO_2 emissions. Unidirectional causality was found between energy use and CO_2 emissions, as well as between CO_2 emissions and GDP; however, no causality was observed from HDI to CO_2 emissions. Robustness checks using FMOLS and DOLS confirmed the significance of HDI and energy consumption at the 5% level. However, the study has limitations, including the failure to account for factors such as political instability, corruption, and institutional influence. Future research should explore these variables and apply the model to other developing countries to inform policies on HDI, economic growth, energy use, and CO_2 emissions.

REFERENCES

Acaravci, A., Ozturk, I. (2010). Electricity consumption-growth nexus: Evidence from panel data for transition countries. *Energy Econ., 32*(3), 604-608.
[http://dx.doi.org/10.1016/j.eneco.2009.10.016]

Alkhathlan, K., Javid, M. (2013). Energy consumption, carbon emissions and economic growth in Saudi Arabia: An aggregate and disaggregate analysis. *Energy Policy, 62*, 1525-1532.
[http://dx.doi.org/10.1016/j.enpol.2013.07.068]

Jahangir Alam, M., Ara Begum, I., Buysse, J., Van Huylenbroeck, G. (2012). Energy consumption, carbon emissions and economic growth nexus in Bangladesh: Cointegration and dynamic causality analysis. *Energy Policy, 45*, 217-225.
[http://dx.doi.org/10.1016/j.enpol.2012.02.022]

Asongu, S.A., Nwachukwu, J.C. (2016). Revolution empirics: predicting the Arab Spring. *Empir. Econ., 51*(2), 439-482.
[http://dx.doi.org/10.1007/s00181-015-1013-0]

Bedir, S., Yilmaz, V.M. (2016). CO_2 emissions and human development in OECD countries: granger causality analysis with a panel data approach. *Eurasian Economic Review, 6*(1), 97-110.
[http://dx.doi.org/10.1007/s40822-015-0037-2]

Brown, R.L., Durbin, J., Evans, J.M. (1975). Techniques for testing the constancy of regression relationships over time. *J. R. Stat. Soc. Series B Stat. Methodol., 37*(2), 149-163.
[http://dx.doi.org/10.1111/j.2517-6161.1975.tb01532.x]

Yavuz, N.Ç. (2014). CO_2 emission, energy consumption, and economic growth for Turkey: evidence from a cointegration test with a structural break. *Energy Sources B Econ. Plan. Policy, 9*(3), 229-235.
[http://dx.doi.org/10.1080/15567249.2011.567222]

Costa, L., Rybski, D., Kropp, J.P. (2011). A human development framework for CO_2 reductions. *PLoS One, 6*(12), e29262.
[http://dx.doi.org/10.1371/journal.pone.0029262] [PMID: 22216227]

Pîrlogea, C. (2012). The human development relies on energy. Panel data evidence. *Procedia Econ. Finance, 3*, 496-501.
[http://dx.doi.org/10.1016/S2212-5671(12)00186-4]

Dickey, D.A., Fuller, W.A. (1981). Likelihood ratio statistics for autoregressive time series with a unit root. *Econometrica, 49*(4), 1057-1072.
[http://dx.doi.org/10.2307/1912517]

Farhani, S., Rejeb, J.B. (2012). Energy consumption, economic growth and CO_2 emissions: Evidence from panel data for MENA region. *International Journal of Energy Economics and Policy, 2*(2), 71-81.

Granger, C.W.J. (1969). Investigating causal relations by econometric models and cross-spectral methods. *Econometrica, 37*(3), 424-438.
[http://dx.doi.org/10.2307/1912791]

Hossain, M.A., Chen, S. (2020). Decoupling of energy-related CO_2 emissions from economic growth: a case study of Bangladesh. *Environ. Sci. Pollut. Res. Int., 27*(17), 20844-20860.
[http://dx.doi.org/10.1007/s11356-020-08541-6] [PMID: 32248420]

Hossain, M.A., Engo, J., Chen, S. (2021). The main factors behind Cameroon's CO_2 emissions before, during and after the economic crisis of the 1980s. *Environ. Dev. Sustain., 23*(3), 4500-4520.
[http://dx.doi.org/10.1007/s10668-020-00785-z]

Hossain, M.A., Chen, S., Khan, A.G. (2021). Decomposition study of energy-related CO_2 emissions from Bangladesh's transport sector development. *Environ. Sci. Pollut. Res. Int., 28*(4), 4676-4690.
[http://dx.doi.org/10.1007/s11356-020-10846-5] [PMID: 32946051]

Islam, S., Khan, M.Z.R. (2017). A review of energy sector of Bangladesh. *Energy Procedia, 110*, 611-618.
[http://dx.doi.org/10.1016/j.egypro.2017.03.193]

Khan, N.H., Ju, Y., Hassan, S.T. (2018). Modeling the impact of economic growth and terrorism on the human development index: collecting evidence from Pakistan. *Environ. Sci. Pollut. Res. Int., 25*(34), 34661-34673.
[http://dx.doi.org/10.1007/s11356-018-3275-5] [PMID: 30324364]

Khan, A.G., Hossain, M.A., Chen, S. (2021). Do financial development, trade openness, economic development, and energy consumption affect carbon emissions for an emerging country? *Environ. Sci. Pollut. Res. Int., 28*(31), 42150-42160.
[http://dx.doi.org/10.1007/s11356-021-13339-1] [PMID: 33797049]

Lee, C.C. (2005). Energy consumption and GDP in developing countries: A cointegrated panel analysis. *Energy Econ., 27*(3), 415-427.
[http://dx.doi.org/10.1016/j.eneco.2005.03.003]

Lean, H.H., Smyth, R. (2010). CO_2 emissions, electricity consumption and output in ASEAN. *Appl. Energy, 87*(6), 1858-1864.
[http://dx.doi.org/10.1016/j.apenergy.2010.02.003]

Martínez, D.M., Ebenhack, B.W. (2008). Understanding the role of energy consumption in human development through the use of saturation phenomena. *Energy Policy, 36*(4), 1430-1435.
[http://dx.doi.org/10.1016/j.enpol.2007.12.016]

Mahmood, M. T., Shahab, S. (2014). Energy, emissions and the economy: Empirical analysis from Pakistan. *Pakistan Development Review, 53*(4, part II), 383-401.
[http://dx.doi.org/10.30541/v53i4IIpp.383-401]

Mehrara, M. (2007). Energy consumption and economic growth: The case of oil exporting countries. *Energy Policy, 35*(5), 2939-2945.
[http://dx.doi.org/10.1016/j.enpol.2006.10.018]

Narayan, P.K., Mishra, S., Narayan, S. (2011). Do market capitalization and stocks traded converge? New global evidence. *J. Bank. Finance, 35*(10), 2771-2781.
[http://dx.doi.org/10.1016/j.jbankfin.2011.03.010]

Narayan, P.K. (2005). The saving and investment nexus for China: evidence from cointegration tests. *Appl. Econ., 37*(17), 1979-1990.
[http://dx.doi.org/10.1080/00036840500278103]

Niu, S., Jia, Y., Wang, W., He, R., Hu, L., Liu, Y. (2013). Electricity consumption and human development level: A comparative analysis based on panel data for 50 countries. *Int. J. Electr. Power Energy Syst., 53*, 338-347.
[http://dx.doi.org/10.1016/j.ijepes.2013.05.024]

Ouedraogo, N.S. (2013). Energy consumption and human development: Evidence from a panel cointegration and error correction model. *Energy, 63*, 28-41.
[http://dx.doi.org/10.1016/j.energy.2013.09.067]

Ozcan, B. (2013). The nexus between carbon emissions, energy consumption and economic growth in Middle East countries: A panel data analysis. *Energy Policy, 62*, 1138-1147.
[http://dx.doi.org/10.1016/j.enpol.2013.07.016]

Pesaran, M.H., Shin, Y., Smith, R.J. (2001). Bounds testing approaches to the analysis of level relationships. *J. Appl. Econ., 16*(3), 289-326.
[http://dx.doi.org/10.1002/jae.616]

Pesaran, M.H., Pesaran, B. (1997). *Working with Microfit4.0: Interactive Econometic Analysis.* Oxford University Press..

Phillips, P. C., Perron, P. (1988). Testing for a unit root in time series regression. *Biometrika, 75*(2), 335-346.
[http://dx.doi.org/10.1093/biomet/75.2.335]

Rahman, M.M., Mamun, S.A.K. (2016). Energy use, international trade and economic growth nexus in Australia: New evidence from an extended growth model. *Renew. Sustain. Energy Rev., 64*, 806-816.
[http://dx.doi.org/10.1016/j.rser.2016.06.039]

Bano, S., Zhao, Y., Ahmad, A., Wang, S., Liu, Y. (2018). Identifying the impacts of human capital on carbon emissions in Pakistan. *J. Clean. Prod., 183*, 1082-1092.
[http://dx.doi.org/10.1016/j.jclepro.2018.02.008]

Sarkar, M.S.K., Sadeka, S., Sikdar, M.M.H., Zaman, B. (2015). Energy consumption and CO_2 emissions: trends and policy implications. *Asia Pacific Journal of Energy and Environment, 2*(3), 220-227.

Salim, R., Yao, Y., Chen, G.S. (2017). Does human capital matter for energy consumption in China? *Energy Econ., 67*, 49-59.
[http://dx.doi.org/10.1016/j.eneco.2017.05.016]

Saboori, B., Sulaiman, J. (2013). Environmental degradation, economic growth and energy consumption: Evidence of the environmental Kuznets curve in Malaysia. *Energy Policy, 60*, 892-905.
[http://dx.doi.org/10.1016/j.enpol.2013.05.099]

Shahbaz, M., Lean, H.H. (2012). *The dynamics of electricity consumption and economic growth: a revisit study of their causality in Pakistan.*
[http://dx.doi.org/10.1016/j.energy.2012.01.048]

Shahbaz, M., Hye, Q.M.A., Tiwari, A.K., Leitão, N.C. (2013). Economic growth, energy consumption, financial development, international trade and CO_2 emissions in Indonesia. *Renew. Sustain. Energy Rev., 25*, 109-121.
[http://dx.doi.org/10.1016/j.rser.2013.04.009]

Shahbaz, M., Mahalik, M.K., Shah, S.H., Sato, J.R. (2016). Time-varying analysis of CO_2 emissions, energy consumption, and economic growth nexus: Statistical experience in next 11 countries. *Energy Policy, 98*, 33-48.
[http://dx.doi.org/10.1016/j.enpol.2016.08.011]

Siddique, H. M. A., Majeed, M. T., Ahmad, H. K. (2016). The impact of urbanization and energy consumption on CO□ emissions in South Asia. *A Research Journal Of South Asian Studies, 31*(2), 745-757.

Steinberger, J.K., Roberts, J.T. (2010). From constraint to sufficiency: The decoupling of energy and carbon from human needs, 1975–2005. *Ecol. Econ., 70*(2), 425-433.
[http://dx.doi.org/10.1016/j.ecolecon.2010.09.014]

Soytas, U., Sari, R., Ewing, B.T. (2007). Energy consumption, income, and carbon emissions in the United States. *Ecol. Econ., 62*(3-4), 482-489.
[http://dx.doi.org/10.1016/j.ecolecon.2006.07.009]

Toda, H.Y., Yamamoto, T. (1995). Statistical inference in vector autoregressions with possibly integrated processes. *J. Econom., 66*(1-2), 225-250.
[http://dx.doi.org/10.1016/0304-4076(94)01616-8]

Tran, N. V., Tran, Q. V., Do, L. T. T., Dinh, L. H., Do, H. T. T. (2019). Trade-off between environment, energy consumption and human development: Do levels of economic development matter? *Energy, 173*, 483-493.

[http://dx.doi.org/10.1016/j.energy.2019.02.072]

United Nations Development Programme, (2014). *Human Development Report 2014: Sustaining Human Progress – Reducing Vulnerabilities And Building Resilience.* Oxford University Press.

United Nations Development Programme, (2020). Human development indices and indicators: 2018 statistical update – Bangladesh. *Human Development Indices And Indicators: 2018 Statistical Update – Bangladesh.* Available form: http://hdr.undp.org/sites/all/themes/hdr_theme/country-notes/BGD.pdf

Wang, Z., Bui, Q., Zhang, B. (2020). The relationship between biomass energy consumption and human development: Empirical evidence from BRICS countries. *Energy, 194*, 116906. [http://dx.doi.org/10.1016/j.energy.2020.116906]

Wang, Z., Danish, Zhang, B., Wang, B. (2018). Renewable energy consumption, economic growth and human development index in Pakistan: Evidence form simultaneous equation model. *J. Clean. Prod., 184*, 1081-1090. [http://dx.doi.org/10.1016/j.jclepro.2018.02.260]

CHAPTER 8

An Overview of Pollution of Drinking Water Due to Unsafe Disposal of Industrial Waste in Raipur, Chhattisgarh

Ghanshyam Shakar[1,*], **Srishti Shaifali Minz**[2] and **Shweta Dewangan**[3]

[1] *School of Engineering & IT, MATS University, Raipur, Chhattisgarh, India*

[2] *Govt. Rajmohini Devi Girls P.G. College, Sarguja University, Ambikapur, Chhattisgarh, India*

[3] *Faculty of Commerce, ICFAI University, Raipur, Chhattisgarh, India*

Abstract: Humanity depends on water, which is the most abundant element on Earth. Groundwater is the most crucial water source for meeting people's daily requirements. As a result, there is now greater dependence on groundwater, which contributes to its overexploitation. As far as groundwater is concerned, several cities in India have reached critically low levels. Due to increased demand and a lack of groundwater replenishment, the groundwater supply is declining. Despite overuse, people have even changed the natural recharge system in the name of development. The current waste management system is insufficient to handle residential and chemical waste. Industrial waste is generally discharged to the ground, alas, and rivers without treatment. This waste is mixed with groundwater, which pollutes it. The dependency on groundwater and the discharge of chemical waste are currently polluting the groundwater, which will become a significant issue in the future. The current study examines the detrimental effects of industrial waste in the Siltara industrial area of Raipur, the state capital of Chhattisgarh, and its surrounding localities. Thirty locations were selected for sampling water, and five samples were collected from each location within the study area of Siltara, which is polluted by industrial effluents. The writers examined the physical and chemical characteristics of the samples collected. After sample collection and analysis, it is determined that the water is highly contaminated and unsafe for human consumption, and several steps should be taken to purify it.

Keywords: Contamination, Groundwater, Industrial effluents, Over-exploitation, Recharge.

INTRODUCTION

The atmospheric environment is a physical and biotic habitat, which one can see, hear, touch, smell, and taste. The synergistic correlation between humans and the

* **Corresponding author Ghanshyam Shakar**: School of Engineering & IT, MATS University, Raipur, Chhattisgarh; E-mail: ghanshyam.geo@gmail.com

environment is essential; the balance between humans and the environment must be maintained at all costs (Agrawal & Singh, 2005). Regrettably, due to a range of human activities, the composition and character of the atmosphere have been severely altered.

Several activities, such as construction, industrialization, and transportation, are necessary for societal development. However, these activities contribute to the production of waste, which in turn pollutes our environment and causes groundwater pollution (Shrivastava, 1994).

Urban growth and industrial activities are making the situation worse because:

1. Natural assets are considered free products, but humanity damages this property.
2. The environment is considered a dumping ground where all byproducts are disposed of untreated. Also, it is impossible to dispose of waste material outside the planet (Singh *et.al.*, 2004)

From the above two statements, it is clear that the time has come for us to use natural resources wisely, producing minimal waste. Intelligent management of natural resources and waste management techniques will help protect the Earth from pollution.

On the other hand, groundwater is the only source available below the Earth's surface that can meet human requirements (Mishra & Shrivastava, 2015). Groundwater is the water that occurs below the water table within the Earth. The occurrence of this water depends on the properties of rocks, such as porosity, permeability, transmissivity, and available storage (Mandal *et al.,* 2012). Additionally, it depends on the region's topography, yearly precipitation, and the area's need, use, and administration.

Industrial Effluents and Wastewater

The word 'affluent' originated from the word effluent, which means to flow out. In other words, effluent refers to the water flow in a river or channel. However, at present, the meaning of effluent has changed to the flow of waste to the land, air, or any other water body. Wastes such as liquids, smoke, and other waste are termed effluents. Effluent is broadly classified into two categories: industrial effluent and municipal effluent. Industrial effluent comes from a specific industry or factory.

The discharges from various companies are known as industrial effluents, and different organic contaminants have been discovered in diverse water resources.

Consequently, industrial effluent causes a significant environmental issue. When released into rivers and lakes, the water quality is affected. Similarly, Table **1** describes the general types of water pollutants found and their significance on aquatic wildlife, *etc.*

Table 1. General types of water pollutants and their significance.

Class of Pollutant	Significance
Heavy metals	Health, toxicity, aquatic biota
Organically bound metals and metalloids	Toxicity, aquatic biota
Inorganic species	Water quality, toxicity, aquatic biota
Trace organic pollutants	Toxicity
Polychlorinated biphenyls	Toxicity, aquatic biota, wildlife
Pesticides	Toxicity, aquatic biota, wildlife
Detergents	Toxicity, aquatic biota
Chemical carcinogens	Incidence of cancer

Raw industrial effluents can be categorized as organic or mineral, depending on the type of pollution that predominates. These effluents may have a high concentration of organic compounds because of a small number of significant pollutants or many molecules with very low concentrations. Table **2**, therefore, describes the types of industrial waste found in water. It clearly shows the type of industrial wastewater, its characteristics, along with examples of the industries it comes from.

Table 2. Types of industrial waste water.

<table>
<tr><th>Type</th><th>Characteristics</th><th>Industry (example)</th></tr>
<tr><td>Organic</td><td>High organic pollution, readily biodegradable</td><td>Food industries</td></tr>
<tr><td rowspan="2">Organic</td><td rowspan="2">High organic pollution, not readily biodegradable</td><td>Refineries</td></tr>
<tr><td>Petrochemical Industries</td></tr>
<tr><td>Organic</td><td>High organic pollution, non-biodegradable</td><td>Organic synthesis</td></tr>
<tr><td rowspan="2">Mineral</td><td>Low organic pollution Toxics</td><td rowspan="2">Steel industries, Electroplating industries, Extractive and mining-based industries</td></tr>
<tr><td>High suspended solids</td></tr>
<tr><td>Miscellaneous</td><td>Organic pollution (with major pollutants), high salinity</td><td>Chemical Industries</td></tr>
<tr><td>Miscellaneous</td><td>Organic pollution (mixture), salinity</td><td>Pulp and paper industries</td></tr>
</table>

Pollution and Water Pollution are Defined

Using polluted groundwater can have a detrimental impact on human health and potentially lead to the spread of various diseases. The pollutant creates a plume within the aquifer. The underground movement of water distributes the contaminants to a greater extent within the aquifer. Its growing edge, sometimes called a plume edge, may cross over into groundwater wells and ponds along with the surface water, such as groundwater springs; this makes the groundwater supplies harmful to humans and wildlife (World Health Organization, 2017).

Although the relationships between groundwater and surface waters are complex, the relationship between groundwater contamination and surface waters has emerged as a significant consequence (Patel *et.al.*, 2016). It is currently the subject of extensive research. For example, groundwater is the source of water in many rivers and lakes. This means that damage to groundwater aquifers could affect the rivers and lakes that depend on groundwater.

Study Area

The city of Raipur, situated within the Raipur district, is located in the Indian state of Chhattisgarh. It serves as the central government for the state of Chhattisgarh, which is situated in that region. 12,461.9 square kilometres of land make up the Raipur district. Within the heart of the state of Chhattisgarh. The topographical sheets No. 64 G/11 and No. 64 G/12 (1:50,000 scale).

The four districts bordering Raipur are: Baloda Bazar district in the north, Durg district in the west, Raigarh district in the east, and the south. In the middle of a vast plain commonly referred to as the "Rice Bowl of India," where wide, distinct varieties of rice are grown, the city of Raipur may be found. The Mahanadi River flows East of Raipur's city limits; its southern bank is covered in thick forest. The Chhota Nagpur Plateau, which stretches northeast through Jharkhand, rises in the north and merges with the Maikal Hills north of Raipur. The Deccan Plateau is south of Raipur.

The Siltara's geocoordinates are latitude 21.3811556 and longitude 81.6637765. Maandhar, which is located 6.4 kilometres away from Siltara, is the railway station closest to the city. The nearest airport to Siltara is Swami Vivekananda Airport, 23.1 kilometers away. Approximately ₹ 716 crore has been invested in this Centre. More than 48 industries have been established, providing direct employment to over 2,772 people. The Siltara Phase 1 is somewhat developed, while development activities are ongoing in Siltara Phase 2.

Geomorphology of the Study Area

The geomorphology reveals that the structure and formation of surface landforms significantly influence the occurrence and movement of groundwater. Remote sensing serves as an advanced and effective tool for identifying surface features such as drainage patterns, lineaments, and landforms, which help in delineating potential groundwater zones. By integrating geomorphological analysis with remote sensing data, researchers can accurately identify areas with high groundwater potential, thereby improving water resource management and planning, particularly in regions with limited direct hydrogeological data.

Geology of the Area

Regional Geology

The 36,000 square km Chhattisgarh Basin unconformably rests on the Bastar Craton's granite-gneiss and volcanic basement. While the western and northern margins of the basin are fault-bounded, the southern and eastern margins have depositional interaction with the rock in the basement.

The NNE-SSW and E-W to ENE-WSW faults define the basin's northern and western boundaries. Broad open warps and large-scale faults that run in a NNW-SSE, NNE-SSW, E-W, and NE-SW direction describe the deformation in the eastern part of the basin. These warps and faults are evident in the strata. One school of thought suggests that these large-scale faults are responsible for revitalizing basin-opening faults, as noted in a 2003 publication by Chaudhuri *et al.*

The Chhattisgarh Supergroup represents a Proterozoic sedimentary sequence in central India. It consists of a well-defined stratigraphic succession, divided into the lower Chandarpur Group and the upper Raipur Group. These formations include limestone, shale, sandstone, and stromatolitic carbonate rocks, indicating shallow marine depositional environments and significant tectonic stability during deposition. Table **3** presents the stratigraphic succession of the Chhattisgarh supergroup, including its formation and lithology in both groups.

Local Geology

The local geology of the Siltara region indicates that it is part of the Chandi Formation, as well as the Proterozoic Raipur Group, which is a member of the Chhattisgarh Super Group. Four primary types of rock can be found in this area: limestone, shale, sandstone, and dolerite intrusion. The area under investigation contains various geological features, including shale, stromatolitic dolomitic limestone with sandstone, laterite, and stromatolite dolomitic limestone. Table **4**. Describes the Stratigraphic Succession of the Study Area, its age, supergroup, group, formation, and lithology. This stratigraphic succession highlights Chhattisgarh's long and eventful geological past, from ancient cratonic roots to the deposition of diverse sedimentary and volcanic rocks that form the basis of its natural resource wealth.

Table 3. Stratigraphic succession chhattisgarh supergroup (Das, *et.al.* 1992; 2003).

<table>
<tr><th>Age</th><th>Supergroup</th><th>Group</th><th>Formation</th><th>Lithology</th></tr>
<tr><td>Quaternary</td><td>Recent to sub-recent</td><td>-</td><td>Alluvium and Laterite</td><td>Sand, Silt, Clay, and lateritic Soil</td></tr>
<tr><td rowspan="9">Proterozoic</td><td rowspan="9">Chhattisgarh Supergroup</td><td rowspan="6">Raipur Group</td><td>Maniyariformation</td><td>-</td></tr>
<tr><td>Hirriformation</td><td>-</td></tr>
<tr><td>Taranga formation</td><td>-</td></tr>
<tr><td>Chandiformation</td><td>Limestone, Sandstone, and Shale</td></tr>
<tr><td>Gunderdehi formation</td><td>Shale</td></tr>
<tr><td>Charmuria formation</td><td>Limestone & Shale</td></tr>
<tr><td rowspan="3">Chandrapur Group</td><td>Kanspathar formation</td><td rowspan="3">Sandstone, Siltstone, Shale, and Conglomerate</td></tr>
<tr><td>Choparadih formation</td></tr>
<tr><td>Lohardi formation</td></tr>
<tr><td>Archaean</td><td colspan="4">Basement crystallines - Granite, Gneisses, Granulite, and Amphibolite</td></tr>
</table>

Slope Map of the Study Area

Because it significantly influences groundwater infiltration and runoff, the slope plays a crucial role in groundwater movement and can be used to identify areas with abundant groundwater (Shakar *et al.,* 2021). A modest slope results in greater groundwater recharge due to high infiltration rates and low runoff, whereas a medium to high slope increases surface runoff and reduces groundwater recharge. SRTM data were used to create a slope map within an ArcGIS environment.

Table 4. Stratigraphic succession of the study area (Das, *et.al.* 1992; 2003).

Age	Supergroup	Group	Formation	Lithology
Proterozoic	Chhattisgarh Supergroup	Raipur Group	Chandi Formation	Limestone, Sandstone & Shale

Depth to Water Level of the Study Area

The amount of groundwater used and the equilibrium between recharge and outflow impact the quantity of groundwater that is present. The data on groundwater collected from CGWB Raipur were utilized to create a water level map (Kumar *et al.,* 2013). The results of the same study ranged between 3 and 12 mg. In addition, the map is divided into four groups: 3-6 mbgl, 6–9 mbgl, 9–12 mbgl, and >12 mbgl.

The present investigation focuses on the Siltara Industrial Area, specifically Phases 1 and 2, as these zones are extensively developed in terms of industrial establishments. This region was chosen due to concerns raised in previous research studies and anecdotal reports from residents, indicating that the groundwater in this area may be unfit for human consumption. Numerous health-related issues have been observed among the population residing nearby, further emphasizing the need for a thorough scientific examination of water quality in the area.

Sample Collection

Groundwater samples were systematically collected in January and February 2020 to assess water quality. A total of 30 distinct locations across the industrial area were selected for sampling. The water samples were gathered in clean, high-grade polythene bottles to ensure that no external contaminants interfered with the analysis. Before collection, borewells at each site were allowed to pump water continuously for at least five minutes, ensuring that stagnant water was removed and that samples reflected the actual groundwater condition. From each selected location, five representative samples were collected to enhance the reliability and accuracy of the findings. The collected samples were securely sealed, appropriately labeled, and preserved using standard procedures. To maintain the integrity of the water chemistry, all samples were immediately frozen after collection. Subsequently, they were transported to a National Accreditation Board for Testing and Calibration Laboratories (NABL)-accredited laboratory (Ogwo *et al.,* 2014), where detailed physicochemical and toxicological analyses were conducted following standard testing protocols. Table **5** below describes the sampling locations, ages, and depths of the Siltara industrial area, as indicated on the map.

Table 5. Sampling Location.

S.N.	Location	Coordinates	Age (Yrs)	Depth (Feet)
1	L1	21°22'17" N, 81°40'20" E	10	110
2	L2	21°22'14" N, 81°40'20" E	10	100
3	L3	21°22'6" N, 81°40'23" E	20	30
4	L4	21°22'13" N, 81°40'35" E	35	150
5	L5	21°22'13" N, 81°40'36" E	90	40
6	L6	21°22'7" N, 81°40'14" E	17	200
7	L7	21°22'17" N, 81°40'5" E	15	205
8	L8	21°22'26" N, 81°40'12" E	6	200
9	L9	21°22'3" N, 81°40'32" E	20	250
10	L10	21°22'14" N, 81°40'45" E	6	65
11	L11	21°21'47" N, 81°39'42" E	20	215
12	L12	21°22'1" N, 81°38'55" E	1	324
13	L13	21°22'18" N, 81°38'39" E	30	306
14	L14	21°21'37" N, 81°37'46" E	10	207
15	L15	21°22'19" N, 81°39'9" E	10	297
16	L16	21°21'34" N, 81°39'7" E	17	195
17	L17	21°22'26" N, 81°39'26" E	14	90
18	L18	21°23'6" N, 81°39'44" E	25	320
19	L19	21°22'50" N, 81°39'39" E	23	78
20	L20	21°15'47" N, 81°39'35" E	65	240
21	L21	21°16'7" N, 81°40'9" E	41	210
22	L22	21°16'5" N, 81°40'19" E	8	315
23	L23	21°16'6" N, 81°40'27" E	30	121
24	L24	21°16'13" N, 81°40'41" E	48	355
25	L25	21°16'15" N, 81°40'47" E	21	290
26	L26	21°15'59" N, 81°40'46" E	8	136
27	L27	21°15'38" N, 81°40'18" E	12	80
28	L28	21°15'19" N, 81°39'58" E	31	167
29	L29	21°15'11" N, 81°39'52" E	26	375
30	L30	21°22'1" N, 81°40'16" E	10	56

RESULTS

The study's results involved a detailed analysis of the physical and chemical parameters of groundwater samples, as presented in Tables **6** and **7**. These specific sampling points will enhance the study's interpretability, reproducibility, and value to other researchers and policymakers, who will also evaluate it in accordance with the standards set by the World Health Organization (WHO). This assessment helped determine the suitability of groundwater for human consumption in the Siltara Industrial Area, Raipur. A comparative summary of the findings is presented in Tables **8** and **9**, providing a clear view of parameter variations across sampling sites. Additionally, a correlation matrix was developed to identify relationships between various ions and contaminants detected in the groundwater (UNICEF, 2019). To better understand the hydrochemical facies and geochemical evolution, both Piper and Durov plots were employed. These graphical tools enabled the classification of water types and the identification of the dominant chemical processes affecting groundwater quality in the region. The combination of these analytical approaches contributed to drawing comprehensive and data-supported conclusions about the extent of groundwater contamination in the study area.

Table 6. Analysis of physical parameters.

S.N.	Location	T (°C)	pH	EC (µs)	TDS (mg/l)	RP (MV)	TH (mg/l)	Alkalinity (mg/l)	Mg+ (mg/l)
1	L1	26.4	6.3	1374	683	250.4	235	730	18.36
2	L2	25.9	6.24	1308	652	237.7	275	720	17.32
3	L3	25.5	6.58	1747	872	215.3	230	850	30.58
4	L4	25.4	6.26	1366	681	222.1	245	650	12.9
5	L5	25.5	6.24	1439	717	245.8	250	730	17.06
6	L6	25.3	6.25	1041	518	253.9	255	550	10.56
7	L7	25.3	6.45	707	352	246.5	110	610	15.5
8	L8	25.3	6.01	1222	611	285.5	165	720	22.78
9	L9	25.3	6.12	1208	602	272.3	225	600	14.98
10	L10	25.4	6.09	874	437	274.1	265	600	15.76
11	L11	25.5	6.32	1021	509	259.9	170	650	17.32
12	L12	25.4	6.22	453	226	274.5	180	750	12.38
13	L13	25.7	6.13	932	463	274.5	230	850	26.94
14	L14	25.5	6.34	1011	509	262.4	200	680	17.84
15	L15	26	6.13	1667	831	264.3	345	850	17.84

(Table 6) cont.....

S.N.	Location	T (°C)	pH	EC (µs)	TDS (mg/l)	RP (MV)	TH (mg/l)	Alkalinity (mg/l)	Mg+ (mg/l)
16	L16	26.1	6.26	1959	987	212.4	255	650	25.86
17	L17	25.2	6	168	221	261.8	425	690	16.84
18	L18	25.7	6.48	568	367	236.5	227	525	15.16
19	L19	25.1	6.12	1256	551	238.5	202	490	21.96
20	L20	26.3	6.21	1330	732	284.6	195	830	9.56
21	L21	25.5	6.33	560	661	211.5	440	670	3.16
22	L22	25.3	6.17	974	507	265.2	130	560	8.56
23	L23	25.6	6.11	1725	670	235.8	180	550	14.59
24	L24	25.5	6.59	1352	315	242.9	358	850	13.12
25	L25	25.9	6.34	912	489	209.7	120	740	28.49
26	L26	26.1	6.23	633	562	225.4	230	690	16.21
27	L27	26.4	6.29	1060	366.65	246.3	250	850	30.42
28	L28	25.3	6.18	759	270	251.9	360	710	28.65
29	L29	25.4	6.42	680	485	213.6	345	490	21.39
30	L30	25.7	6.37	1502	601	248.2	140	610	5.68

Table 7. Analysis of chemical parameters.

S.N.	Location	NO3	PO4	F	Cl	Na	Ca	K	SO4	NH4
1	L1	6.485	0.51	4.9	896	75.61	16.11	1.59	60.15	16.2
2	L2	9.023	0.567	4.66	854	71.33	15.1	1.59	54.37	16.2
3	L3	19.95	0.87	7.4	874	125.9	17.62	12.46	100.1	14.3
4	L4	20.08	0.567	3.8	836	53.12	15.1	41.91	100.5	17.4
5	L5	19.98	0.784	3.26	754	70.26	17.62	47.94	101.8	15.1
6	L6	13.95	0.822	3.2	658	43.49	13.08	0.25	123.6	15.8
7	L7	13.53	0.728	4.64	516	63.83	13.59	5.76	39.94	11.9
8	L8	20.18	0.605	3.12	680	93.81	15.1	17.81	60.56	17.2
9	L9	19.58	0.558	2.72	738	61.69	15.6	1.08	59.32	12.6
10	L10	20.01	0.605	3.24	726	64.9	17.11	2.41	122	15.5
11	L11	4.416	0.605	3.12	658	71.33	10.06	0.25	92.32	11.8
12	L12	8.813	0.567	4.06	472	50.98	16.11	2.26	91.08	8.16
13	L13	19.57	0.614	4.18	668	116.9	15.1	1.59	58.09	10.1
14	L14	18.95	0.662	4.12	762	73.47	14.09	0.41	93.15	13.4
15	L15	19.93	0.614	4.8	1018	73.47	24.66	64.67	171.9	11.1
16	L16	5.864	0.513	2.86	296	37.52	36.65	3.75	63.22	9.63

(Table 7) cont.....

S.N.	Location	NO3	PO4	F	Cl	Na	Ca	K	SO4	NH4
17	L17	17.26	0.741	3.15	1250	44.13	10.35	7.61	86.41	21.5
18	L18	14.12	0.834	5.63	678	120.16	12.14	42.21	58.16	17.2
19	L19	16.33	0.654	4.77	482	51.85	16.63	15.63	100.2	14.9
20	L20	19.54	0.615	7.12	310	75.33	18.62	2.23	79.35	9.12
21	L21	18.15	0.608	6.32	1313	85.42	10.12	62.75	152.7	15.7
22	L22	18.47	0.869	3.57	286	110.6	14.52	8.21	170.6	19.1
23	L23	7.325	0.647	4.51	455	116.9	19.45	1.12	51.02	13.2
24	L24	12.31	0.529	6.48	980	42.75	25.16	54.18	89.55	10.5
25	L25	11.59	0.657	5.77	825	86.14	24.31	7.26	79.44	14.6
26	L26	15.37	0.781	4.13	1096	91.21	31.25	7.06	63.33	18.5
27	L27	17.86	0.702	3.62	1369	88.96	21.39	1.03	93.18	16.4
28	L28	13.85	0.834	7.25	857	115.38	18.6	0.38	120.2	10.2
29	L29	9.671	0.522	2.96	1214	121.3	19.41	52.76	105.1	9.61
30	L30	5.225	0.656	5.38	1160	46.21	17.96	1.96	56.12	22

Table 8. Comparison between standard and observed values.

S.N.	Species	WHO Prescriptions, (mg/l)	Observed Value, (Low toHighvalue,mg/l)	Average Value
1.	Temperature(°C)	7 –11	25 -26.4	25.6033
2.	pH	6.5 -8.5	6.01 -6.68	6.275
3.	EC	500	4.53 –1973	1056.551
4.	TDS	600	2.26 –985	527.508
5.	RP	-	208.7 -285.5	251.87
6.	TH	100 -500	110 –450	227.5
7.	Mg^{+}	30	2.5 -30.58	14.68
8.	Alkalinity	300	490 –850	653.33
9.	$-NO_3$	45	4.416 -20.45	16.98
10.	$3-PO_4$	5	0.51 -0.87	0.616
11.	F	1.5	2.58 -7.4	3.528
12.	Cl	250	280 –1404	749.4
13.	Na^{+}	20	10.3 -125.94	60.45
14.	Ca^{2+}	75	9.57 -36.75	18.04

(Table 8) cont.....

S.N.	Species	WHO Prescriptions, (mg/l)	Observed Value, (Low toHighvalue,mg/l)	Average Value
15.	K^{+}	25	0.25 -64.68	7.60
16.	SO_4^{2-}	200	5.70 -171.93	73.33
17.	NH_4^{+}	-	8.16 -22.30	13.30

WHO Prescriptions Against Observed Value of Physical and Chemical Parameters

The WHO Guidelines for Drinking-water Quality (GDWQ) offer a comprehensive framework, not rigid "standards." They emphasize health-based targets for a vast range of physical, chemical, and microbial parameters. Comparing your observed groundwater values directly with these guidelines reveals exceedances, highlighting potential public health risks. This comparison is crucial for identifying contaminants that require intervention (*e.g.*, treatment, alternative sources) to ensure the water is safe for consumption.

Correlation Matrix of Ions for Siltara

A correlation matrix of ions for groundwater in Siltara, Raipur, would typically show the statistical relationships between the concentrations of various dissolved ions (*e.g.*, Ca^{2+}, Mg^{2+}, Na^{+}, K^{+}, HCO_3^{-}, Cl^{-}, SO_4^{2-}, NO_3^{-}, F^{-}). In Table **8**, the standard and observed values are shown. A correlation matrix presents a table where each cell contains the **correlation coefficient** (usually Pearson's r) between two variables.

Durov Plot

The Durov diagram is a powerful graphical tool used in hydrogeochemistry to visualize the chemical composition of groundwater samples and infer the geochemical processes controlling their chemistry. It is a composite plot that combines two ternary diagrams into a single plot. This is because the Durov diagram plots the milliequivalent percentages of cations against the percentages of negative ions (Yadav *et al.,* 2018). The sides of the structure come together to form a binary plot with a centre rectangular shape that compares the total cation concentrations to the total anion concentrations. It is more likely that water samples with comparable characteristics will be plotted together as a grouping.

Here is a diagram illustrating the similarities and differences between groundwater samples.

Table 9. Correlation matrix.

	T°C	pH	EC (µs)	TDS (mg/l)	RP (MV)	TH	Alkalinity	No_3^-	PO_4^{3-}	F^-	Cl^-	Na^+	Mg^{2+}	Ca^+	K^+	SO_4^{2+}	NH_4^+
T°C	1.00	-	-	-	-	-	-	-	-	-	-	-	-	-	-	-	-
pH	0.34	1.00	-	-	-	-	-	-	-	-	-	-	-	-	-	-	-
EC (µs)	0.23	-0.10	1.00	-	-	-	-	-	-	-	-	-	-	-	-	-	-
TDS (mg/l)	0.23	-0.10	1.00	1.00	-	-	-	-	-	-	-	-	-	-	-	-	-
RP(MV)	-0.21	-0.43	-0.43	-0.43	1.00	-	-	-	-	-	-	-	-	-	-	-	-
TH	0.29	-0.14	0.72	0.72	-0.35	1.00	-	-	-	-	-	-	-	-	-	-	-
Alkalinity	0.20	0.00	0.11	0.10	0.16	-0.15	1.00	-	-	-	-	-	-	-	-	-	-
No_3^-	-0.22	-0.26	0.28	0.28	-0.03	0.19	-0.03	1.00	-	-	-	-	-	-	-	-	-
PO_4^{3-}	-0.28	0.33	0.18	0.18	-0.19	-0.07	0.23	0.11	1.00	-	-	-	-	-	-	-	-
F^-	0.14	0.27	0.23	0.22	-0.18	0.02	0.65	-0.20	0.46	1.00	-	-	-	-	-	-	-
Cl^-	0.29	-0.08	0.88	0.88	-0.45	0.83	-0.10	0.20	-0.06	0.04	1.00	-	-	-	-	-	-
Na^+	0.04	-0.11	0.59	0.59	-0.11	0.21	0.61	0.13	0.34	0.55	0.41	1.00	-	-	-	-	-
Mg^{2+}	0.04	-0.11	0.59	0.59	-0.11	0.21	0.61	0.13	0.34	0.55	0.41	1	1.00	-	-	-	-
Ca^+	0.18	-0.12	0.61	0.61	-0.33	0.83	-0.24	0.36	-0.18	-0.14	0.80	0.13	0.13	1.00	-	-	-
K^+	0.04	-0.17	0.39	0.39	-0.04	0.22	0.44	0.25	0.22	0.26	0.18	0.20	0.20	0.09	1.00	-	-
SO_4^{2+}	0.10	-0.08	0.51	0.51	-0.12	0.47	0.31	0.16	0.33	0.20	0.47	0.44	0.44	0.26	0.52	1.00	-
NH_4^+	0.05	-0.17	0.43	0.43	-0.35	0.32	-0.11	0.17	0.05	0.01	0.25	0.24	0.24	0.15	0.11	0.28	1.00

Piper Plot

The Piper (1944) triliner diagram can be used to explain the geochemistry of groundwater. In the current investigation, these diagrams were created using Grapher-16. Piper diagrams are multidimensional plots that display the miliequivalent (meq) percentage concentrations of major cations (Ca 2+, Mg2+, Na+, and K+) and anions (HCO3-, SO4 2-, and Cl-) in two triangle fields. The fields are then projected farther into the central diamond region. Piper diagrams analyse the relationship between the anions and the significant cations. It is more likely that water samples with comparable characteristics will be plotted together as a grouping.

CONCLUSION

The current study's findings reveal that the existing status of groundwater in the Siltara Industrial Area, R Aipurur, is not suitable for drinking purposes, except at a few sites. Furthermore, the hydrochemical analysis indicates that the situation may continue to deteriorate in the future as well. As a result of the extremely high percentage of water tested, it is clear that the desirable limits according to WHO criteria are being approached, and the water is dangerously close to exceeding the maximum level permitted. On the other hand, the safe disposal of industrial wastewater and effluents has become a challenge for industrial managers, researchers, and scientists. For this reason, it is essential to have suitable effluent treatment techniques and disposal methods that are appropriate for purifying the groundwater and preventing further deterioration.

Significant leaching of various chemical components into the groundwater has been seen in both regions. This has resulted in the enrichment of many anions and cations, which indicates that sources from outside the area may have contaminated the groundwater. It is now time to update the original meaning of "water supply management" to "water demand management." This revision is now necessary. To improve groundwater quality in India, efficient wastewater treatment and solid waste management should be the most critical components. This should be done rather than offering a range of benefits.

REFERENCES

Agrawal, M., Singh, B. (2005). Impact of coal mining on water quality of Damodar River – A case study of Jharia coalfield, India. *Int. J. Environ. Res. Dev., 1*(2), 121-128.

Agrawal, M., Singh, B. (2005). Impact of coal mining on water quality of Damodar River: A case study of Jharia coalfield, India. *International Journal Of Environmental Research And Development* (2)

Banerjee, U.S., Gupta, S. (2013). Impact of industrial waste effluents on river Damodar adjacent to Durgapur industrial complex, West Bengal, India. *Environ. Monit. Assess., 185*(3), 2083-2094.
[http://dx.doi.org/10.1007/s10661-012-2690-1] [PMID: 22623168]

Bartram, J., Balance, R. (1996). *Water quality monitoring: A practical guide to designing and implementing freshwater quality studies and monitoring programmes.* United Nations Environment Programme & World Health Organization.
[http://dx.doi.org/10.4324/9780203476796]

Bhateria, R., Jain, D. (2016). Water quality assessment of lake water: a review. *Sustain. Water Resour. Manag., 2*(2), 161-173.
[http://dx.doi.org/10.1007/s40899-015-0014-7]

Biswas, A., Jana, A., Sharma, S.P. (2012). Delineation of groundwater potential using satellite remote sensing and GIS technique: A case study from Ganjam District, Orissa. *Res. J. Recent Sci., 1*, 59-66.

Bougherira, N., Hania, A., Djabria, L., Toumia, F., Chaffaia, H., Haied, N., Nechema, D., Sedratia, N. (2013). Impact of urban and industrial waste on surface and groundwater in the region of Annaba, Algeria. *International Research Journal Of Environmental Sciences, 2*(4), 34-38.

Bureau of Indian Standards. (2012). *Indian Standard Drinking Water—Specification (2nd Revision) (Is 10500:2012).* Available form: https://www.services.bis.gov.in

Central Ground Water Board. (2010). *Jaipur District Brochure*Government of India.

Chinhanga, R.J. (2016). Impact of industrial effluent from an iron and steel company on the physico-chemical quality of Kiekie River water in Zimbabwe. *Research Journal of Environmental Toxicology, 10*(3), 172-182.

Das, B. (2014). Environmental impact due to iron ore mining in Chhattisgarh. *Recent Research in Science and Technology, 6*(1), 27-29.

Dhembare, A.J. (2013). Analysis of drinking water of 10 blocks of Jalgaon district (MS), India. *Arch. Appl. Sci. Res., 5*(2), 154-158.

Dubey, A.K., Gupta, A. (2015). Groundwater pollution and emerging environmental challenges in industrial areas of Chhattisgarh. *International Journal of Scientific and Research Publications, 5*(3), 1-4.

Elumalai, V., Brindha, K., Lakshmanan, E. (2017). Human exposure risk assessment due to heavy metals in groundwater by pollution index and multivariate statistical methods: A case study from South Africa. *Water, 9*(4), 234.
[http://dx.doi.org/10.3390/w9040234]

Faisal, B.M.R., Majumder, R.K., Mohammed, J.U., Mohammed, A.H. (2013). Studies on heavy metals in industrial effluent, river and groundwater of Savar Industrial Area, Bangladesh by principal component analysis. *Int. Res. J. Environ. Sci., 2*(4), 34-38.

Inayathullah, M., Paul, J.M. (2013). Water quality index assessment of ground water in Jakkur Sub Watershed of Bangalore, Karnataka, India. *Int. J. Civil Struct. Environ. Infrastruct. Eng. Res. Dev., 3*, 99-108.

Iram, S., Kanwal, S., Ahmad, I., Tabassam, T., Suthar, V., Mahmood-ul-Hassan, M. (2013). Assessment of physicochemical parameters of wastewater samples. *Environ. Monit. Assess., 185*(3), 2503-2515.
[http://dx.doi.org/10.1007/s10661-012-2727-5] [PMID: 22832844]

Jose, S. K., Jayasree, R., Santhosh, R., Rajendran, S. (2012). Identification of ground water potential zones in Palakkad District, Kerala, through multi-criteria analysis techniques using geo-information technology. *Bonfring International Journal Of Industrial Engineering And Management Science, 2*(Special Issue 1), 62-68.

Jothivenkatachalam, K., Nithya, A., Chandra Mohan, S. (2010). Correlation analysis of drinking water quality in and around Perur block of Coimbatore District, Tamil Nadu, India. *Rasayan J. Chem., 3*(4), 649-654.

Kumar, P.P., Yadav, S.R., Singh, R.R. (2013). Impact of industrial effluents on ground and surface water quality in Raigarh District of Chhattisgarh, India. *Int. J. Res. Chem. Environ., 3*(3), 209-215.

Kumari, S., Singh, A.K., Verma, A.K., Yaduvanshi, N.P.S. (2014). Assessment of groundwater quality using water quality index in Varanasi District, Uttar Pradesh, India. *Environ. Monit. Assess., 186*(1), 501-514.
[http://dx.doi.org/10.1007/s10661-013-3393-y] [PMID: 23996647]

Mandal, S., Bhattacharya, A.K., Sinha, S. (2012). A study on the contamination of iron in groundwater and its impact in eastern India. *Int. J. Environ. Sci., 3*(1), 252-258.

Mishra, S., Shrivastava, A. (2015). A study on heavy metal contamination in groundwater around industrial area of Korba, Chhattisgarh, India. *Int. J. Environ. Sci., 6*(2), 168-175.

Mohanty, C., Behera, S. C. (2010). Integrated remote sensing and GIS study for hydrogeomorphological mapping and delineation of groundwater potential zones in Khallikote Block, Ganjam District, Orissa. *Journal Of The Indian Society Of Remote Sensing* (2)
[http://dx.doi.org/10.1007/s12524-010-0022-9]

National Green Tribunal. (2017). *Principal Bench Order 09.05.2017 on OA No. 593/2017.* Retrieved from. [https://www.greentribunal.gov.

National Remote Sensing Centre (NRSC). (2010). *Remote sensing applications.* Indian Space Research Organization, Department of Space, Government of India.

Ogwo, P. A., Eze, V. C., Nweze, N. O. (2014). Impact of industrial effluents discharge on the quality of Nwiyi River, Enugu, South Eastern Nigeria. *Iosr Journal Of Environmental Science, Toxicology And Food Technology (Iosr-jestft), 8*(11), 22-27.

Pandey, C., Rajwade, R.P. (2015). Analysis of water quality in different locations of BALCO industrial area of Korba, Chhattisgarh, India. *International Journal of Research in Engineering, Science and Technologies, 1*, 45-49.

Pandey, C., Rajwade, R.P. (2015). The physicochemical characteristics of surface and groundwater of BALCO industrial area of Korba city. *Journal of Civil Engineering and Environmental Technology, 2*, 23-25.

Pandey, R., Singh, D., Kumar, S. (2015). Analysis of groundwater of Raigarh coal mining area. *Int. J. Chem. Stud., 3*(1), 38-40.

Pandey, V., Tiwari, R.K. (2012). Water pollution due to industrial effluents in Raigarh district, Chhattisgarh, India. *J. Environ. Res. Dev., 6*(3), 553-559.

Panigrahi, K., Das, B., Patel, B. (2021). Chemical analysis of surface water of Raipur, Chhattisgarh to evaluate the consequences of industrial effluents, India. Samriddhi: A Journal of Physical Sciences. *Engineering and Technology, 13*(2), 118-124.

Patel, K.S., Yadav, A., Sharma, C. (2016). Heavy metal contamination in industrial region of Chhattisgarh and its impact on human health. *Journal of Environmental Science, Toxicology and Food Technology, 10*(4), 20-24.

Patel, T. H., Reddy, V. V., Miseram, S. R., Kumar, P. S. K. (2017). Impact of iron and steel industry on ground water quality of Tungabhadra River water in Bellary District (India). *International Journal Of Advanced Research, Ideas And Innovations In Technology (Ijariit), 3*(1), 983-985.

Patra, A.K., Jha, A. (2014). Physico-chemical characterization of groundwater in industrial area of Raipur city, Chhattisgarh, India. *Int. J. Environ. Sci., 5*(1), 45-52.

Pradhan, B. (2009). Groundwater potential zonation for basaltic watersheds using satellite remote sensing data and GIS techniques. *Open Geosci., 1*(1), 120-129.
[http://dx.doi.org/10.2478/v10085-009-0008-5]

Ritwik, H., Jha, S., Karn, S.K. (2017). Iron contamination: A major threat in Chhattisgarh, India. *Int. J. Res. Stud. Biosci., 5*(1), 46-48.

Sahu, S.K., Sharma, M. (2016). Water quality assessment of industrial areas in Chhattisgarh using multivariate statistical techniques. *Journal of Environmental Science, Computer Science and Engineering & Technology, 5*(3), 145-152.

Shakar, G., Das, B. (2020). An impact of industrial effluents on groundwater quality of Siltara Industrial Area, Raipur, C.G. *Journal of University of Shanghai for Science and Technology, 22*, 442-456.

Shakar, G., Das, B., Patel, B. (2021). Analysis of groundwater quality using spatial techniques in Naya Raipur Atal Nagar, Chhattisgarh, India. Samriddhi: A *Journal of Physical Sciences, Engineering and Technology, 13*(2), 109-117.

Sharma, D., Kansal, A. (2011). Water quality analysis of River Yamuna using water quality index in the national capital territory, India (2000–2009). *Appl. Water Sci., 1*(3-4), 147-157.
[http://dx.doi.org/10.1007/s13201-011-0011-4]

Sharma, M., Goel, S. (2016). Groundwater contamination by heavy metals in Raipur, Chhattisgarh, India. *Int. J. Sci. Res., 5*(5), 135-137.

Sharma, M., Sahu, B.L. (2014). Groundwater quality assessment for drinking purpose in industrial area of Chhattisgarh using WQI and GIS. *Int. J. Environ. Sci., 5*(2), 310-317.

Sharma, S.K., Sharma, M.P., Malviya, S. (2013). Evaluation of water quality of Narmada River with reference to physico-chemical parameters at Hoshangabad city, MP, India. *Research Journal of Chemical Sciences, 3*(1), 40-48.

Shrivastava, A., Kumar, R. (2013). Groundwater quality assessment in industrial area of Bhilai, Chhattisgarh, India. *International Journal Of Scientific And Research Publications, 3*(7), 1-5.

Singh, G., Singh, D. D., Sharma, S. K. (2013). Effect of polluted surface water on groundwater: A case study of Budha Nullah. *Iosr Journal Of Mechanical And Civil Engineering, 5*(5), 1-8.

Singh, K.P., Malik, A., Mohan, D., Sinha, S. (2004). Multivariate statistical techniques for the evaluation of spatial and temporal variations in water quality of Gomti River (India)—a case study. *Water Res., 38*(18), 3980-3992.
[http://dx.doi.org/10.1016/j.watres.2004.06.011] [PMID: 15380988]

Singh, S.K., Singh, D. (2014). Impact of industrial effluents on groundwater quality in Raipur region, Chhattisgarh, India. *Int. J. Environ. Sci., 4*(3), 175-182.

Singh, V.K., Tiwari, T.N. (2010). Water quality index for assessment of water quality in Ganga River. *Int. J. Environ. Sci., 2*(2), 29-35.

Sinha, A.K., Sinha, M. (2015). Assessment of physicochemical characteristics of groundwater in industrial areas of Chhattisgarh, India. *International Journal of Scientific and Research Publications, 5*(6), 1-4.

Sinha, A., Gupta, R. (2015). Assessment of drinking water quality in Durg District, Chhattisgarh, India. *Int. J. Eng. Res. Appl., 5*(3), 50-54.

Tiwari, M.K., Dwivedi, A. (2016). Impact of iron and steel industries on groundwater quality in Central India. *Pollut. Res., 35*(3), 585-591.

Tiwari, M.K., Bajpai, S., Dewangan, U.K. (2016). Air and leaching pollution scenario by iron and steel plants in Central India. *Elixir Pollution, 91*, 43495-43501.

UNICEF. (2019). *Water, sanitation and hygiene.*https://www.unicef.org/india/what-we-do/water-sanitat-on-hygiene

Vaishnav, M.M., Dewangan, S. (2012). Analytical and statistical evaluation of surface and sub- surface water of BALCO industrial area, Korba, C.G., India. *Int. J. Environ. Sci., 2*(3), 1369-1379.
[http://dx.doi.org/10.6088/ijes.00202030023]

Verma, S., Yadav, R. (2013). Assessment of heavy metal contamination in groundwater in industrial region of Raigarh, Chhattisgarh. *Int. J. Environ. Eng., 5*(1), 14-21.

Verma, T.K., Tiwari, K.L., Jadhav, S.K. (2014). Correlation between iron pollution and physicochemical characteristics of effluent of steel industries from Urla, Raipur (Chhattisgarh), India. *Journal of Sustainable Environmental Research, 3*(2), 183-186.

Verma, T.K., Tiwari, K.L., Jadhav, S.K. (2014). Physicochemical studies of steel industries effluent of Urla, Raipur (C.G.) about iron contamination, biotechnology and traditional knowledge. *Journal of Sustainable Environmental Research, 3*(2), 265-270.

World Health Organization. (2017). *Guidelines For Drinking-water Quality: Fourth Edition Incorporating The First Addendum.* WHO. Available form: https://www.who.int/publications/i/item/9789241549950

Yadav, A., Kumar, A., Singh, G. (2015). Water quality assessment of groundwater resources in industrial area of Chhattisgarh, India. *Int. J. Curr. Res., 7*(10), 21465-21469.

Yadav, A., Rajhans, K.P., Ramteke, S., Sahu, B.L., Patel, K.S., Blazhev, B. (2016). Contamination of industrial wastewater in Central India. *J. Environ. Prot. (Irvine Calif.), 7*(1), 72-81.
[http://dx.doi.org/10.4236/jep.2016.71007]

Yadav, K.K., Gupta, N., Kumar, V., Khan, S.A., Kumar, A., Bishnoi, N.R., Kumar, S. (2018). A review of emerging treatment technologies for industrial wastewater. *Environ. Sci. Pollut. Res. Int., 25*(31), 30860-30881.
[PMID: 30209766]

CHAPTER 9

Sustainable Urban Development: Managing Construction and Demolition Waste Methods in the Indian Context

Srishti Shaifali Minz[1,*], **Shweta Dewangan**[2], **Ghanshyam Shakar**[3] and **Sanjeev Kumar**[4]

[1] *Govt. Rajmohini Devi Girls P.G. College, Sarguja University, Ambikapur, Chhattisgarh, India*

[2] *Faculty of Commerce, ICFAI University, Raipur, Chhattisgarh, India*

[3] *School of Engineering & IT, MATS University, Raipur, Chhattisgarh, India*

[4] *School of Hotel Management & Tourism, Lovely Professional University Phagwara, Punjab, India*

Abstract: The rapid growth of cities and towns, as well as infrastructure booms in India, has significantly escalated the future of Construction and Demolition waste, posing substantial environmental and public health challenges. Currently, the waste management practices are inadequate, resulting in improper disposal, resource wastage, and ecological degradation. This paper examines sustainable solutions for managing Construction and Demolition (C&D) waste in urban India, with a focus on waste minimization, recycling, policy enforcement, and the adoption of innovative technologies. The study examines the arrangement and volume of C&D waste and offers a critical evaluation of existing policies, including the 'Construction and Demolition Waste Management Rules, 2016'. The paper highlights successful case studies in global practices and their applicability in India. The challenges are identified and discussed with sustainable solutions. Technological innovations, including innovative crushers, automated sorting systems, and the integration of AI and IoT for efficient waste monitoring, are explored. The role of governance in enhancing compliance and policy, as well as incentivising private sector participation, is emphasised, explaining the importance of community engagement and public awareness campaigns. The paper concludes with the need for a coordinated effort from all stakeholders to implement sustainable Construction and Demolition (C&D) waste management practices that directly contribute to urban sustainability and resource conservation.

Keywords: Construction and Demolition, Ecological Degradation, Smart Waste Management Systems, Modular Construction, Waste Audits.

* **Corresponding author Srishti Shaifali Minz:** Govt. Rajmohini Devi Girls P.G. College, Sarguja University, Ambikapur, Chhattisgarh, India; E-mail: drsrishtisminz@gmail.com

Sanjeev Kumar, Mohammad Badruddoza Talukder & A.K. Haghi (Eds.)

INTRODUCTION

Overview of Urbanization Trends and Their Impact on 'C&D Waste Generation in India

India has been experiencing unpredictable growth in urban areas. According to the UN, the urban population in India is projected to reach 600 million by 2030, accounting for approximately 40% of the total population. This rapid urbanisation is driven by natural population growth and migration from rural to urban areas in search of better employment opportunities and living conditions. The economic liberalization of the 1990s accelerated urban development, resulting in substantial investments in infrastructure projects, including housing, transportation, public utilities, and commercial complexes. This urban growth has led to extensive construction activities across the nation.

The 'Smart Cities Mission' was launched in 2015. Jain and Tiwari (2019) state that the aim is to promote sustainable and inclusive development by improving infrastructure and services in 100 selected cities. This initiative has encouraged construction activities like multiplexes, buildings, roads, and other urban infrastructure. The growing urban population has driven a high demand for housing and real estate. Government initiatives, such as the Housing for All by 2022 program, aim to construct affordable housing units for citizens who directly contribute to construction activities.

Construction and Demolition Waste Management Rules, 2016

Scope and Applicability

This rule applies to all waste generated from the construction, remodeling, repair, and demolition of any civil structure in urban and rural areas. This includes building materials, debris, and rubble resulting from the construction, remodelling, repair, and demolition of any civil structure of an individual, organisation, or authority that generates Construction and Demolition waste.

Responsibilities of 'Waste Generators'

'Waste generators' are individuals, builders, and government bodies. They are responsible for:

- **Segregation**: Ensuring that C&D waste is segregated into recyclable and non-recyclable components before disposal.
- **Storage**: Waste is stored in accordance with the guidelines of local authorities.

- **Transportation**: Ensures safe transportation of waste to designated processing or disposal sites.
- **Disposal**: Disposing of C&D waste at only authorised collection, processing, or disposal sites.
- **Informing Authorities**: Provide information to the local authority regarding the production of Construction and Demolition (C&D) waste.

Duties of Service Providers and Contractors

The Service providers such as local municipal authorities, PWD, and contractors must perform these duties of Removal and Disposal, Record Keeping, and Compliance, which means removing C&D waste promptly and ensuring that it isdisposedof at the designated place, maintaining records of C&D waste production, transportation, and disposal of and ensuring compliance with the norms and standards set by local authorities for managing waste.

Duties of Local Authorities

Local authorities are tasked with Authorization, Monitoring, Planning and Implementation, Infrastructure Development, and Public Awareness, which defines providing authorisation to C&D waste processing facilities, ensuring compliance with waste management rules by all stakeholders, developing waste management plans, establishing collection, transportation, and processing facilities, and conducting awareness programs for waste generators regarding proper waste management practices.

Duties of State Government

State governments are responsible for formulating policies and strategies for C&D waste management within their state, conducting training programs, and monitoring and supervising the implementation of these rules by local authorities and other stakeholders.

Duties of the Central Pollution Control Board (CPCB)

The CPCB is responsible for formulating guidelines for the environmentally sound management of Construction and Demolition (C&D) waste, promoting research and development (R&D) activities for the recycling and reuse of C&D waste, and coordinating with state pollution control boards for the implementation of these guidelines.

Duties of State Pollution Control Boards (SPCBs)

SPCBs must monitor compliance with the rules and ensure that local authorities adhere to the established rules. They are also responsible for setting environmental standards for C&D waste processing and disposal facilities, as well as approving and certifying waste processing and recycling facilities.

Waste Processing and Recycling Facilities

These facilities obtain authorization from the local authority to comply with environmental standards and utilize new technologies, thereby minimizing environmental impact, maximizing resource reclamation, and operating together to process and recycle C&D waste efficiently.

Use of Recycled Materials

Government bodies must promote the use of recycled materials in construction activities and develop standards and specifications to support this effort.

Monitoring and Reporting

Local authorities submit annual compliance reports to the SPCBs, and SPCBs and CPCBs conduct inspections periodically to ensure adherence to the rules.

The Construction and Demolition Waste Management Rules, 2016, provide a comprehensive framework for managing Construction and Demolition waste in India. Maheshwari and Gupta (2017) emphasize the importance of segregation, recycling, and proper disposal in reducing environmental impact and promoting resource efficiency. Effective implementation of these rules requires coordinated efforts from all stakeholders.

Impact on C&D Waste Generation

The construction activities have resulted in a substantial increase in the volume of Construction and Demolition (C&D) waste. India is projected to generate approximately 530 million tonnes of Construction and Demolition (C&D) waste annually, comprising concrete, bricks, metals, wood, glass, and plastics. Urban renewal projects, including the demolition of old structures and renovation activities, have contributed to the growth of a modernized city and its redevelopment. Theredevelopment. The demolition of outdated buildings produces large quantities of C&D waste.

A significant portion of Construction and Demolition (C&D) waste is not properly managed. It is often disposed of in unauthorised areas, leading to illegal dumping and environmental degradation. This improper disposal contributes to soil and water pollution, obstructs drainage systems, and causes urban flooding. The generation of C&D waste depletes natural resources, as new construction requires fresh raw materials. Furthermore, the manufacturing and disposal of these materials harm the environment through increased energy consumption, habitat destruction, and the release of greenhouse gases. The current infrastructure for processing and recycling is insufficient, according to Jain and Tiwari (2019), despite the great potential for recycling and reusing C&D waste materials. Difficult situations might be responsible for the lack of standard procedures and recycling incentives. Consequently, the ecological footprint of construction activities has increased, and valuable natural resources are being depleted. Although the Construction and Demolition Waste Management Rules of 2016 were introduced to tackle these issues, local municipalities and states still inconsistently apply them. The effective management of C&D waste is further obstructed by weak enforcement and failure to comply with these regulations.

Research by Sharma and Panwar (2019) and Singh and Mahesh (2018) indicates that this situation presents both opportunities and challenges for Solid Waste Management (SWM). To tackle the issue, a comprehensive strategy is required, which includes promoting waste reduction methods, increasing public awareness, enhancing recycling facilities, as examined by Gupta and Goel (2021), and strengthening the enforcement of policies. By adopting sustainable practices and leveraging technological advancements, India can enhance its management of Construction and Demolition (C&D) waste, thereby aiding environmental conservation and promoting urban sustainability.

Challenges Associated with the Unregulated Disposal of C&D Waste

The waste generated from Construction and Demolition activities in urban regions of India poses a significant challenge in terms of environmental, social, and economic impacts. The unregulated disposal of this waste leads to various critical issues and problems:

1. **Environmental Harm**
   a. **Soil & Water Contamination:** As a result of uncontrolled waste disposal, harmful substances such as asbestos, lead, and other chemicals infiltrate the soil. The local vegetation and wildlife face threats when this waste is dumped in open spaces since it leaches hazardous elements into the soil, diminishing its fertility. Rainwater can transport these pollutants into water bodies, leading to water contamination that endangers drinking water sources and affects aquatic ecosystems.

b. **Air Pollution:** Inadequate waste management and disposal contribute to air pollution through dust and particulate matter. Demolition dust contains hazardous materials, including silica, which can lead to silicosis and other severe respiratory conditions. Furthermore, burning waste outdoors releases harmful gases that worsen air quality and contribute to climate change.
c. **Loss of Biodiversity:** Illegal waste dumping in wetlands, forests, and other ecologically significant areas leads to habitat destruction and decreased ecological diversity. The degradation of these sites can have a severe impact on wildlife and is crucial for maintaining environmental balance.

2. **Public Health Risks**
   a. **Disease and Health Hazards:** The indiscriminate disposal of Construction and Demolition waste creates breeding grounds for disease vectors like mosquitoes and rodents, facilitating the spread of illnesses such as malaria, dengue, and leptospirosis. Exposure to hazardous materials leads to various health issues, including respiratory ailments, skin irritations, and chronic conditions like cancer.
   b. **Workplace Risks:** Workers who manage and dispose of Construction and Demolition waste without proper safety equipment and training face significant dangers of accidents and health complications. Exposure to toxic substances, heavy items, and sharp materials increases the likelihood of occupational diseases and injuries.
3. **Urban Aesthetic and Infrastructure Challenges**
   a. **Visual Disarray:** Waste discarded in public spaces, along roadsides, and in empty lots results in unattractive scenery and visual clutter. The presence of waste heaps diminishes the visual appeal of urban environments, adversely affecting tourism and local businesses.
   b. **Damage to Infrastructure:** The illegal disposal of waste can obstruct drainage systems, leading to water accumulation and urban flooding during heavy rains. This inconveniences residents and damages public infrastructure, including roads, bridges, and buildings.
4. Resource Inefficiency
   a. **Waste of Recyclable Materials:** The unregulated disposal of Construction and Demolition (C&D) waste hinders the recovery and recycling of valuable resources like bricks, metals, concrete, and wood. These materials could be repurposed for new construction projects, thereby reducing reliance on raw materials and minimizing environmental impact.
   b. **Increased Pressure on Landfills:** The improper handling of C&D waste places additional strain on existing landfill sites, hastening their depletion and prompting new landfill development. This entails financial investment and further exacerbates land scarcity and environmental harm.

5. **Economic Implications**
   a. **Costs of Cleanup and Remediation:** The illegal disposal of C&D waste incurs significant cleanup and remediation expenses for local municipal authorities. These costs could be avoided or reduced through effective waste management practices. With adequate funding, the department could redirect its focus toward other essential services.
   b. **Missed Economic Opportunities:** Ineffective implementation of C&D waste management systems restricts the potential for generating economic opportunities in the recycling and waste management fields. A well-regulated C&D waste recycling sector can create jobs and stimulate economic growth.

Objectives

Successful Construction and Demolition (C&D) waste management is essential for sustainable urban development in India. The objectives for this study are outlined as follows:

- To identify and analyse to mitigate environmental impacts, conserve resources, and improve public health.
- To address the challenges involved with unregulated disposal and the coordinated efforts from all stakeholders
- To implement global practice solutions in India.

With this, India can move towards a more resource-efficient and environmentally responsible urban development model.

LITERATURE REVIEW

C&D Waste in India

The Ministry of Environment, Forest and Climate Change (2016) states that primary sources of C&D waste include construction, renovation, and demolition activities in both residential and commercial sectors. According to a Central Pollution Control Board (CPCB) study, C&D waste comprises concrete, bricks, metals, wood, glass, plastics, and other materials used in construction. The composition of waste includes: concrete and rubble (40-50%), Bricks and masonry (30-40%), Metals (5-10%), Wood (5-10%), and Miscellaneous Materials (including glass, plastics, *etc.*) (1-5%).

Chandra (2016) estimates India generates approximately 530 million tonnes of C&D waste annually. This substantial volume poses challenges such as collection, transportation, processing, and disposal, which require effective management strategies.

The principal sources of C&D waste are:

- **Construction activities** include new construction, renovations, and repairs to residential, commercial, and industrial buildings.
- **Demolition activities** include tearing down old buildings and structures to make way for new developments.
- **Infrastructure projects** encompass the construction and maintenance of roads, bridges, flyovers, and other public infrastructure.
- **Natural disasters**, such as earthquakes and floods, result in significant structural damage and debris.

The volume of Construction and Demolition (C&D) waste generation in India is substantial and has been increasing rapidly due to urbanization and infrastructure development. However, accurate data collection and reporting remain challenging due to inadequate monitoring and enforcement mechanisms.

Regulations and Policies

The Ministry of Environment, Forests, and Climate Change introduced the Construction and Demolition Waste Management Rules in 2016 to address the growing challenges associated with Construction and Demolition waste. These regulations emphasize the establishment of recycling facilities, the implementation of proper collection and transportation methods, and the source segregation of waste. The Urban Swachh Bharat Mission emphasizes the importance of effective waste management, particularly regarding Construction and Demolition (C&D) waste, to promote clean and sustainable urban environments (MoEFCC, 2016). The National Urban Sanitation Policy aims to improve sanitation infrastructure through enhanced waste management and urban planning, thereby positively influencing the management of Construction and Demolition waste. However, due to a lack of awareness, inadequate infrastructure, and enforcement challenges, these regulations are often applied inconsistently by governments and municipalities.

Global Best Practices

Nations such as Germany and Japan have established efficient C&D Waste Management Systems that can serve as benchmarks for India. These systems are characterized by stringent regulations, advanced technologies, and high recycling rates, employing measures such as Extended Producer Responsibility (EPR), the

Construction Material Recycling Law, Green Building Practices, and Public Awareness initiatives (Fischer & Davidsen, 2017).

CURRENT CHALLENGES IN MANAGING C&D WASTE IN URBAN INDIA

The management of Construction and Demolition waste in urban India faces numerous obstacles that hinder the successful implementation of sustainable waste management strategies. These obstacles include:

Insufficient Awareness and Training among Stakeholders

A significant issue in managing Construction and Demolition waste is the lack of knowledge and training among numerous stakeholders, including builders, contractors, municipal staff, and the public. Many stakeholders are unaware of the best practices for waste segregation, recycling, and reuse, as well as the adverse effects of improper Construction and Demolition (C&D) waste disposal on the environment and public health.

Weak Enforcement of Existing Policies

The Construction and Demolition Waste Management Rules, 2016, offer a comprehensive framework for managing C&D waste. However, these regulations are not adequately enforced. Local authorities often lack the necessary resources and capacity to monitor and ensure compliance effectively. As a result, illegal disposal of C&D waste continues to be widespread, worsening environmental and health challenges.

Limited Recycling Infrastructure and Technologies

India's recycling infrastructure for C&D waste remains underdeveloped. There are too few recycling facilities, and existing ones typically operate below capacity due to technological and logistical issues. The adoption of advanced recycling technologies, such as mobile crushers and automated sorting systems, remains minimal, further hindering the effective processing and recycling of Construction and Demolition (C&D) waste.

Ineffective Segregation and Collection Systems

Effective C&D waste management begins with proper segregation at the source. However, urban areas in India currently have poorly designed and ineffective systems for segregation and collection. The mixed nature of waste streams complicates the recovery of recyclable materials, resulting in more waste being sent to landfills. Additionally, the lack of authorised collection sites and transportation infrastructure further complicates waste management efforts.

Land Scarcity for Waste Disposal and Recycling Facilities

Urban regions in India face considerable land shortages, which make it challenging to identify appropriate locations for waste disposal and recycling facilities. The high land cost and competing demands for space in densely populated cities worsen this issue. This constraint hinders the development of essential infrastructure for C&D waste management, leading to an increase in illegal dumping in unauthorized locations.

SUSTAINABLE SOLUTIONS FOR C&D WASTE MANAGEMENT

Waste Reduction at the Source: Reducing waste at its origin is vital to minimising the overall ecological consequences of Construction and Demolition (C&D) activities. This practice involves designing and executing methods that reduce waste production during the construction phase, thereby decreasing the amount of waste that requires management, recycling, or disposal at a later stage. In urban India, where rapid urban growth and infrastructure expansion significantly contribute to waste generation, implementing source reduction techniques can greatly enhance sustainability. A fundamental approach to mitigating the environmental effects of Construction and Demolition waste in urban India is waste reduction at the source. The construction industry can significantly reduce waste output by focusing on deconstruction-friendly design, optimising material use, modular construction, conducting waste audits, and promoting on-site sorting. Modular construction minimises environmental impact by reducing the need for heavy machinery at the job site. As Kumar and Agrawal (2020) noted, the optimal use of materials decreases waste and lowers material costs, fostering more sustainable building practices. Effective planning and waste audits ensure reduced waste generation on-site, cut costs, and improve resource efficiency. On-site sorting enhances waste management practices, decreases contamination rates, and increases the value of recyclable materials. Nevertheless, addressing awareness, costs, regulations, and infrastructure obstacles is essential for broadly implementing these practices. Through education, policy enforcement, and investments in infrastructure, waste reduction can become a pivotal element of sustainable urban development in India.

Recycling and Reuse of C&D Waste: The recycling and repurposing of C&D waste are crucial for diminishing the environmental impact of the construction sector and supporting a circular economy. By employing recycling and reuse methods, it is feasible to mitigate ecological impacts, conserve resources, and reduce the volume of Construction and Demolition (C&D) waste sent to landfills.

Technologies for Recycling C&D Waste

a. **Mechanical Crushing and Screening:** Crushers can mechanically break down waste materials into smaller fragments, such as concrete and masonry. These fragments are then screened to separate the fine aggregates from the coarse aggregates that can be repurposed in new construction.
b. **Mobile Recycling Machines:** Mobile recycling units can be directly transported to construction sites to process C&D waste on location. This practice diminishes the ecological impact of transporting waste to distant recycling facilities and reduces transport expenses.
c. **Advanced Sorting Technologies:** Metals, wood, polymers, and concrete are among the materials that are separated using automated sorting technologies such as air classification, vibrating screens, and magnets. These technologies increase recycling efficiency by guaranteeing improved purity and quality of recovered materials.
d. **Cold Recycling for Asphalt**: Asphalt pavements can be cold-recycled using milling machines that break down the existing asphalt and reuse it in new pavement mixes, reducing the need for virgin asphalt.
e. **Composting of Organic Waste**: Although not a significant component of C&D waste, organic materials (such as plants and trees from demolition sites) can be composted and utilized for landscaping or agricultural purposes.

Importance of Recycling and Reuse

a. **Environmental Benefits:** Reusing and recycling Construction and Demolition debris reduces the demand for natural raw materials, thereby preserving valuable resources such as metals, aggregates, and lumber. Recycling reduces greenhouse gas emissions from raw material extraction, processing, and transportation. Recycling properly reduces the burden on landfills and the harm that unmanaged Construction and Demolition (C&D) waste disposal causes to the environment.
b. **Financial Gains:** The cost of raw materials for building projects can be significantly reduced by recycling, reusing materials, and utilizing Construction and Demolition debris. Using recycled materials reduces disposal expenses because less waste is sent to landfills. Moreover, recycling enhances the local economy by creating job opportunities in collecting, processing, and selling reclaimed materials.
c. **Social Benefits:** Recycling initiatives can foster a more environmentally aware society by raising public understanding of sustainable waste management practices. By engaging communities, recycling programs can promote social responsibility and collective actions to protect the environment.

Despite the numerous opportunities to recycle and repurpose Construction and Demolition (C&D) materials, significant challenges exist due to regulatory hurdles, high contamination levels, and inadequate infrastructure. To fully harness the potential of recycling Construction and Demolition waste, advancements in recycling technology, a more favourable regulatory framework, and increased demand for recycled materials are essential. With a stronger and more coordinated approach to recycling and reuse, India can move towards more sustainable building practices by minimising waste and conserving vital resources.

Technological Innovations

To more effectively, sustainably, and profitably manage Construction and Demolition (C&D) waste, technological progress is crucial. Enhancements in waste processing techniques, material recovery systems, and recycling technology can significantly reduce the environmental footprint of C&D activities. Implementing advanced technologies is vital for ensuring adequate waste management in urban India.

Automated Sorting Systems

Automated sorting technologies have revolutionised the recycling of Construction and Demolition debris by improving the efficiency and accuracy of waste separation. These systems utilise mechanical, optical, and air classification methods to differentiate various materials, such as concrete, metals, plastics, and wood, from mixed waste streams.

Optical sorting machines utilize cameras and sensors to distinguish between different materials based on their composition, color, and shape. These resources are then automatically sorted into separate streams. Air classifiers separate lighter materials, such as paper, plastics, and wood, from heavier items, like metals and concrete, by using airflow to distinguish components based on their density. Magnets extract steel and iron, two types of ferrous metals, from Construction and Demolition (C&D) waste.

These technologies aid in enhancing recovery rates and the quality of recycled materials, reducing the need for manual labour and minimising human error, improving efficiency, and recycling mixed waste streams that were previously difficult to sort.

Mobile Recycling Units

Mobile recycling units enable Construction and Demolition (C&D) waste to be processed directly at construction or demolition sites, reducing transportation costs and minimizing the environmental impact of hauling waste over long distances. These units typically include portable crushers, screens, and separation equipment that efficiently process materials like concrete, wood, and metals on-site.

Mobile crushers crush concrete and masonry waste into smaller, reusable aggregates used in road construction, pavements, or as sub-base material for new buildings. Mobile Screening plants can separate finer aggregates from coarser materials, allowing for the production of higher-quality recycled materials.

These technologies help reduce the need for transporting waste to distant recycling facilities, thereby cutting down on carbon emissions and minimizing the environmental impact of C&D waste. This is achieved by promoting on-site processing, reducing landfill use, and enhancing the cost-effectiveness of recycling by eliminating transportation and handling costs.

Recycled Concrete Aggregates (RCA) Processing Technologies

Waste concrete from demolished buildings, roads, or other infrastructure is crushed and processed to create Recycled Concrete Aggregates (RCA). Technologies for RCA processing have advanced to enhance the quality of recycled aggregate, qualifying it for use in structural applications.

High-Pressure Water Jetting: This technique improves the quality of RCA by cleaning the aggregates and eliminating contaminants such as dust, dirt, and coatings using high-pressure water jets. The size and grading of the aggregates can be precisely controlled by modern crushers and screens, ensuring that the recycled concrete meets industry requirements for use in new construction projects.

By providing a sustainable alternative to natural aggregates, these technologies reduce the pressure on quarries and the extraction of natural resources while also improving the quality of recycled concrete aggregates (RCA) for a broader range of construction uses, including structural concrete, and minimising the environmental impact of concrete production by repurposing existing materials.

3D Printing with Recycled Materials

As 3D printing technology gains popularity in the construction industry, it also facilitates the sustainable disposal of Construction and Demolition debris. Researchers are exploring the use of recycled materials, such as glass, plastics, and crushed concrete, as feedstock for 3D printing building materials. Utilising locally sourced Construction and Demolition waste, this method can be applied to create road infrastructure, building components, and even entire buildings.

There are two types of 3D printing: concrete and plastic. 3D printers can produce beams, columns, and walls using concrete mixed with recycled aggregates and eco-friendly materials. Plastics retrieved from Construction and Demolition waste can be processed into 3D printing filaments for fabricating components, fixtures, and fittings in the construction sector.

According to ICON 3D Printing (2021), these technologies foster innovation in construction techniques, enabling more sustainable and efficient building practices while helping decrease material waste by repurposing Construction and Demolition debris into new building components. Furthermore, utilising locally available waste materials may reduce construction costs and energy use.

Biodegradable Building Materials and Waste Minimization

A significant advancement in managing Construction and Demolition waste is the development of biodegradable building materials sourced from organic or industrial waste. Alam and Hossain (2020) describe these materials, such as composites made from mushrooms or bio-bricks formed from agricultural byproducts, as eco-friendly alternatives to conventional building materials.

Mushroom-Based Materials: Mycelium composites and other fungus-derived materials can be cultivated and moulded into construction elements, offering a biodegradable option to traditional foams and plastics in building projects.

As noted by Orozco and Basart (2021), these technologies provide a sustainable method for managing organic waste by transforming it into functional building materials, thereby lowering the carbon footprint of the construction industry and promoting the reuse of waste materials for new construction, all of which align with the principles of a circular economy.

In addition to enhancing the efficacy and quality of recycling processes, these technologies bolster the overall sustainability of the construction industry. However, to fully reap the benefits of these innovations, India must invest in infrastructure, implement robust regulatory frameworks, and promote collaboration across the industry to create a circular economy for Construction and Demolition waste.

Policy and Governance: Addressing the challenges of managing Construction and Demolition (C&D) waste in urban India requires robust governance and policy structures. These structures ensure adherence to regulations, guide the implementation of waste management practices, and offer incentives for eco-friendly solutions. The existing governance frameworks and legislation in India aim to reduce waste generation, promote recycling, and improve overall waste management processes in the construction industry.

a. **Construction and Demolition Waste Management Rules, 2016:** These regulations obligate construction firms to segregate, recycle, and reuse waste management systems. To mitigate the environmental impact of waste disposal, developers and contractors must ensure that a significant portion of Construction and Demolition debris is recycled or repurposed. Local authorities monitor compliance, ensure proper waste handling, and establish C&D waste management facilities.
b. **Extended Producer Responsibility (EPR):** EPR is essential in encouraging the construction sector to take responsibility for the entire lifecycle of construction materials, from production to disposal. EPR requires that manufacturers and developers facilitate the collection and recycling of building materials. Holding manufacturers accountable for waste management aims to reduce reliance on landfills and promote recycling efforts.
c. **State and Local Regulations:** State and local governments can implement and enforce tailored policies for managing Construction and Demolition waste, considering local conditions. These regulations often cover waste transportation, processing, and segregation, and promote the establishment of recycling facilities in major cities. However, there is usually an enforcement gap due to insufficient resources for compliance and monitoring.
d. **Public-Private Partnerships (PPP):** By leveraging the expertise of both public and private sectors, public-private partnerships (PPP) can significantly enhance the management of Construction and Demolition waste. PPPs can help finance the construction of recycling facilities, improve infrastructure, and provide incentives for recycling and reusing construction debris.
e. **Incentives and Financial Support:** Public-private partnerships can significantly enhance waste management for Construction and Demolition by integrating the strengths of both sectors. PPPs can improve infrastructure, fund the establishment of recycling facilities, and encourage the recycling and reuse of Construction and Demolition waste.

For C&D waste management in India to succeed, strong regulations, effective governance, and active participation from all stakeholders are essential. The C&D Waste Management Rules, 2016, provide a framework for sustainable waste management alongside regional regulations and PPP models. However, the key to success lies in strengthening enforcement, fostering corporate engagement, and investing in infrastructure development.

Community and Stakeholder Engagement: According to TOMRA Sorting Solutions (2021), it is vital for all stakeholders—including government officials, businesses, construction firms, and the community—to actively engage in the effective management of Construction and Demolition (C&D) debris in urban India. Community and stakeholder engagement is essential for raising awareness, ensuring compliance, and encouraging participation in activities related to waste reduction, segregation, recycling, and reuse.

Government and Regulatory Bodies: Government agencies are primarily tasked with the formulation, implementation, and enforcement of regulations such as the Construction and Demolition Waste Management Rules, 2016. These laws mandate the proper disposal, recycling, and segregation of Construction and Demolition waste, with specific targets for reducing waste. Governments must also develop and maintain the infrastructure necessary for C&D waste management, which includes recycling centers, disposal sites, and waste collection systems.

Engagement Strategies

- **Public Consultations:** Engage communities in the decision-making process through public consultations to gain a deeper understanding of local issues, preferences, and challenges related to compliance with waste management regulations.
- **Collaboration with Local Authorities:** Empower local municipal bodies to manage C&D waste at the city level, ensuring that waste management strategies are aligned with the specific needs of the community.

Private Sector (Construction Companies and Recycling Industry): As outlined by Choudhary and Jha (2020), construction companies must comply with waste management regulations, conduct source segregation of waste, and implement waste collection systems. The private sector bears the primary responsibility for developing recycling systems and technologies that enhance the management of Construction and Demolition debris. Furthermore, private enterprises play a critical role in establishing and running recycling facilities.

Engagement Strategies

- **Incentives and Support:** Provide financial incentives, such as tax reductions or subsidies, to encourage construction companies to invest in recycling infrastructure and adopt sustainable practices (Nwokolo *et al.,* 2024).
- **Capacity Building:** Implement training programs for construction workers and project managers on best practices in waste management, including effective

waste segregation, recycling methods, and adherence to environmental standards.

Community and Citizens: Communities play an essential role in enhancing the efficiency of C&D waste management systems by practicing waste segregation at home and within neighborhoods. Additionally, residents can advocate for improved waste management practices and policies. Promoting the reuse of materials from demolished buildings, such as metals, bricks, and wood, can significantly reduce the generation of Construction and Demolition (C&D) waste and lessen the demand for new construction materials.

Engagement Strategies

- **Educational Campaigns:** Launch awareness initiatives and outreach programs that educate citizens on the environmental and economic advantages of proper waste segregation, recycling, and reuse. Schools, local organizations, and media channels can be utilized to disseminate this knowledge.
- **Community-driven Recycling Initiatives:** Motivate neighborhood groups or communities to create small-scale collection and recycling facilities for C&D materials. This can build a sense of shared responsibility and contribute to local waste reduction efforts.

Academia and Research Institutions: Research organizations and academic institutions play a key role in developing new technologies and methodologies for managing Construction and Demolition waste. Innovations in construction practices, advanced recycling techniques, and research on sustainable materials can provide valuable insights for improving waste management systems. To develop knowledge in waste management and sustainable construction methods, universities and training facilities can offer professionals, students, and construction industry workers specialized courses and workshops.

Engagement Strategies

- **Collaborative Research Projects**: Encourage collaboration among academia, industry, and government to research and develop effective strategies for managing Construction and Demolition (C&D) waste. This can include the development of sustainable building materials or the optimization of recycling processes.
- **Educational Programs**: Universities and technical institutions can design curricula and certification programs focused on sustainable construction, waste management, and environmental practices.

NGOs and Civil Society: NGOs play a critical role in advocating for environmental protection and sustainable waste management practices. They have raised awareness about the consequences of reducing Construction and Demolition (C&D) waste and promoting community involvement in recycling initiatives.NGOs can assist in implementing government policies at the grassroots level, facilitating community engagement in waste management programs, and building local capacity for recycling.

Engagement Strategies

- **Public Campaigns and Advocacy**: NGOs can organize campaigns and events to raise awareness about C&D waste management, emphasizing the risks associated with improper disposal of C&D materials to the environment and health.
- **Collaborative Community Programs**: Partner with local communities, municipal authorities, and construction companies to design and implement waste reduction initiatives, such as the collection and recycling of construction debris.

Informal Sector (Waste Pickers and Laborers): In India, waste pickers and informal laborers are crucial in the collection, sorting, and recycling of C&D waste (Rajpal *et al.,* 2024), especially in urban areas. Their contribution, although informal, is significant in the recovery of recyclable materials, such as metal, glass, and plastic (Singh *et al.,* 2021a), from demolition sites.

Engagement Strategies

- **Formalizing the Informal Sector**: Engage the informal sector through training and integration into formal waste management systems, providing them with legal recognition, better working conditions, and a share of the profits from recycling.
- **Social and Financial Support**: Provide financial and social support, such as fair wages, insurance, and access to healthcare, to enhance the working conditions and welfare of waste pickers involved in demolition waste management.

Circular Economy Approach: The Circular Economy (CE) framework represents a sustainable model aimed at minimizing waste (Electricity Distribution, n.d.) and effectively utilizing existing resources by establishing a closed-loop system. In terms of Construction and Demolition (C&D) waste management, it prioritizes the reduction, reuse, and recycling of materials rather than disposing of them in landfills. The CE framework promotes the design of buildings and infrastructure that can be easily dismantled and reused, thereby ensuring the sustainable use of materials. Key strategies involve recycling concrete into aggregates for new construction purposes, reusing demolition debris for roadworks, and selecting sustainable materials. The CE model offers advantages to the construction industry by reducing the demand for raw materials extracted from the natural environment, conserving natural resources, and minimizing environmental impacts. Adopting CE practices in C&D waste management promotes innovation in waste reduction, material recovery, and sustainable construction, thereby advancing the overall objectives of environmental protection and economic efficiency.

CASE STUDIES

Burari C&D Waste Recycling Facility, Delhi: This facility, established in 2017, serves as a key example of how to effectively recycle and manage Construction and Demolition waste in Indian urban areas. Located in the Burari neighborhood of North Delhi, this facility aims to underscore the challenges posed by the massive quantities of waste generated in the city. As one of the pioneering projects in India to adopt a sustainable method for managing building and demolition debris, this initiative converts waste into essential raw materials for construction, such as bricks, sand, and aggregates.

Impact Analysis

1. **Waste Reduction:** The Burari facility has the capacity to process up to 496 tonnes of Construction and Demolition waste daily, which helps minimize the amount of waste sent to landfills, according to the Municipal Corporation of Delhi (2019). By recycling, the facility prevents hundreds of tons of waste from entering landfills each year, thereby alleviating pressure on the existing waste management system.
2. **Material Recovery:** Some of the Construction and Demolition waste is transformed into valuable construction materials, including aggregates for concrete, bricks, and tiles, as reported by IL&FS Environment (2018). This recovery of materials has reduced the need for virgin resources, such as crushed stone and river sand, which have become increasingly scarce and environmentally damaging to extract.
3. **Environmental Advantages:** The recycling of Construction and Demolition waste lessens the environmental impact of construction activities (Chen *et al.*, 2021), such as the depletion of natural resources, energy use, and greenhouse gas emissions associated with the extraction and transportation of raw materials. By reducing the demand for new resources, the facility helps mitigate deforestation and habitat destruction associated with quarrying and mining activities.

4. **Job Creation:** The Burari facility has created job opportunities for various stakeholders, including laborers, truck drivers, and skilled professionals involved in the recycling processes. It has also contributed to the formal inclusion of waste pickers in the recycling effort, benefiting the informal waste-picking sector.

Initiatives for Recycling C&D Waste by the Ahmedabad Municipal Corporation: In the rapidly growing urban areas of India, the Ahmedabad Municipal Corporation (AMC) has taken the initiative to implement systems for managing Construction and Demolition waste. Recognizing the significant issues caused by Construction and Demolition (C&D) waste (Singh *et al.,* 2021b), the AMC launched various programs aimed at addressing the city's challenges related to waste generation, disposal, and recycling. These initiatives are designed to support the circular economy, reduce dependence on landfills, and promote sustainable waste management practices. An assessment of AMC's efforts can provide valuable insights into effective waste management strategies and their broader impacts.

In 2016, AMC established a dedicated C&D Waste Recycling Plant at the Mochampura Waste Management Facility. This facility has the capacity to process approximately 500 tons of Construction and Demolition waste daily and is specifically designed for this purpose. Utilizing advanced technologies, the plant is capable of sorting, crushing, and recycling waste into reusable products, including aggregates, sand, and materials suitable for road construction. AMC (2018) made efforts to ensure that contractors and construction firms separate C&D waste right at the source. The corporation's regulations emphasize that contractors must categorize demolition debris into recyclable and non-recyclable waste and transport only the separated waste to processing facilities, as noted by the Ministry of Housing and Urban Affairs (2019).To inform the public, builders, and construction businesses about the negative effects of waste segregation and the benefits of reusing Construction and Demolition materials for the environment, AMC has initiated awareness campaigns. The goal of public outreach initiatives, including workshops, seminars, and media campaigns, is to increase recycling compliance and participation.

Construction projects must now manage waste in an environmentally friendly manner as a result of the corporation's integration of the Construction and Demolition Waste Management Rules, 2016 (CPCB | Central Pollution Control Board, n.d.) into local rules. The AMC's waste management teams monitor C&D waste disposal procedures, conduct routine inspections, and impose fines for non-compliance as enforcement measures.

Impact Analysis

1. **Decreased Landfill Reliance:** Since the establishment of the C&D waste recycling plant, the amount of construction waste sent to landfills has diminished. Given the limited land availability and growing waste management challenges in the city, AMC has alleviated the need for new landfill locations by transforming waste into valuable building materials. This recycling facility has played a significant role in supporting Solid Waste Management (SWM) principles by diverting a considerable portion of Construction and Demolition waste from landfills.
2. **Resource Recovery & Circular Economy:** A key outcome of AMC's recycling initiatives for Construction and Demolition waste has been the recovery of valuable materials, including aggregates, crushed concrete, and sand (Chen *et al.,* 2021). Utilizing these materials in new construction projects diminishes the need for natural resources such as sand and gravel (Gautam *et al.,* 2024). By adopting the circular economy model within the construction sector, AMC has promoted environmental sustainability and the conservation of resources.
3. **Environmental Benefits:** The recycling of C&D waste mitigates the environmental impact of construction activities, which include raw material extraction, energy use, and carbon emissions. Producing recycled aggregates for construction reduces the need for mining, an activity often associated with deforestation, habitat loss, and soil degradation. Furthermore, incorporating recycled materials into road construction and building projects contributes to a lower carbon footprint associated with the creation of new materials.
4. **Economic Advantages and Job Creation:** AMC's C&D waste recycling facility has created direct and indirect employment opportunities. This includes roles in facility management and operations, as well as jobs in the sorting, processing, and sale of recycled materials. By decreasing reliance on costly imported raw materials, a local market for recycled Construction and Demolition (C&D) materials has emerged, thereby enhancing the local economy. The construction sector benefits from reduced material costs by promoting the use of recycled materials, which can potentially lower overall building expenses.
5. **Compliance and Implementation Challenges:** Despite the advantages of AMC's initiatives, challenges remain in enforcing recycling and waste segregation regulations. Many informal contractors and smaller construction projects struggle to comply with these regulations due to limited infrastructure, resources, and a lack of awareness. To enhance waste management outcomes, AMC must intensify its educational efforts, outreach, and enforcement to ensure compliance across all segments of the construction sector.

6. **Collaboration Between Public and Private Sectors:** The success of AMC's C&D waste recycling initiatives highlights the importance of collaboration between public and private sectors (Rwelamila & Abdul-Aziz, 2020). By partnering with private waste management companies, the organization has leveraged the private sector's technology and expertise, which has been vital to the growth of these programs. Additional collaborations with construction firms can enhance waste reduction practices and encourage the adoption of recycling technologies within the industry.

Global Examples of C&D Waste Management and their Application in India

A major global concern is the efficient management of Construction and Demolition (C&D) waste, especially in nations like India that are rapidly urbanizing (Tata_Realty_ESG_Report_FY_2023-24.Pdf, n.d.). Several nations have developed innovative and successful C&D waste management systems that can offer India valuable lessons. We have examined international best practices for managing Construction and Demolition waste from countries such as Germany, Japan, and the Netherlands, and investigated how these techniques can be applied in the Indian context.

Germany: Advanced Recycling Technologies and Regulatory Frameworks

Germany has the most advanced infrastructure for recycling Construction and Demolition (C&D) waste among all countries worldwide. The country boasts numerous state-of-the-art recycling facilities that process a wide range of Construction and Demolition (C&D) materials into reusable products. The German government enforces stringent regulations that mandate the recycling of Construction and Demolition (C&D) waste. (Federal Ministry for the Environment, Nature Conservation, and Nuclear Safety, 2018) verifies the country's Construction Waste Management Ordinance requires waste producers to segregate materials at the source and ensures that at least 70% of C&D waste is recycled. Germany has developed a robust market for recycled Construction and Demolition (C&D) materials. These products are widely used in construction, helping to reduce the demand for virgin resources.

India can invest in creating cutting-edge C&D waste recycling facilities equipped with modern technology to process a variety of waste types effectively. Stronger adherence to recycling and trash segregation laws can be achieved by implementing laws modelled after those in Germany. Developing a market for recycled Construction and Demolition (C&D) materials through government procurement policies, certifications, and incentives can encourage their use in construction projects.

Japan: Zero-waste and Circular Economy Approaches

Japan's waste management philosophy emphasizes achieving zero waste through the principles of reduce, reuse, and recycle (3Rs). This approach extends to C&D waste, where efforts are made to minimize waste generation and maximize recycling. Japan utilizes advanced technologies for sorting, processing, and recycling Construction and Demolition (C&D) waste. Automated systems ensure high efficiency and quality in material recovery. The Japanese government (2016) states that Japan's success in waste management is partly due to high levels of public awareness and participation. The government conducts extensive campaigns to educate citizens and businesses about the importance of waste segregation and recycling.

Takahashi (2015) displays that Indian cities can adopt Japan's zero-waste philosophy by setting ambitious targets for waste reduction and reutilization in the construction sector. Investing in smart sorting and recycling technologies can enhance the quality and efficiency of Construction and Demolition (C&D) waste management in India. Tam & Tam (2006 explore public awareness campaigns and education programs that can be implemented to inform stakeholders about the benefits and practices of C&D waste management.

The Netherlands: Public-Private Partnerships (PPPs) and Innovative Recycling Techniques Integration of Waste Management and Urban Planning

Public-Private Partnerships (PPPs) have proven effective in the Netherlands for managing Construction and Demolition waste. As stated by the Dutch Ministry of Infrastructure and Water Management (2016), these partnerships establish efficient waste management systems by integrating the capabilities and knowledge of both the public and private sectors. The Netherlands employs advanced techniques to recycle Construction and Demolition materials, such as reusing asphalt in road projects and incorporating recycled concrete into new buildings. The country is recognized for integrating waste management practices into urban planning, ensuring that recycling and waste processing facilities are incorporated into the urban landscape. Key strategies outlined by the Ellen MacArthur Foundation (2015) include Urban Mining, which involves extracting materials from existing urban structures and buildings that are ready for demolition. Additionally, establishing decentralized recycling centers enables the creation of small-scale recycling facilities within urban settings, making them more accessible and reducing transportation expenses.

Implementing PPPs could significantly improve how Indian municipalities handle Construction and Demolition waste. Collaboration with private companies can bring in innovative technologies and operational expertise. By adopting advanced recycling techniques from the Netherlands, such as repurposing old asphalt and concrete, India can enhance resource recovery and promote eco-friendly construction practices. Moreover, embedding waste management into urban planning in India will ensure that recycling facilities are conveniently located and capable of managing the substantial amounts of Construction and Demolition (C&D) waste produced. Promoting urban mining can lead to the recovery of materials from outdated buildings and infrastructure. Additionally, government support for research and development in recycling technologies can foster innovations specifically tailored to India's needs, thereby enhancing the sustainability of its construction industry.

Denmark: Public-Private Partnerships and Circular Economy

Public-Private Partnerships (PPPs) have been successfully utilized in Denmark to handle Construction and Demolition (C&D) waste, incorporating circular economy principles into waste management practices. In conjunction with Circular Economy Models, which focus on designing structures for disassembly and material reuse while also ensuring materials can be recovered and recycled throughout the building's lifecycle, the Danish Environmental Protection Agency (2020) explores the collaborations between municipalities and private firms to develop and operate recycling facilities as a critical practice that offers financial incentives for construction projects that adopt sustainable practices and recycle materials.

India could enhance its C&D waste management system by embracing Denmark's PPP model. Recycling facilities could be created through collaborations between private companies and government entities. Starting with Miljøstyrelsen (2017), principles of the circular economy were encouraged.

Australia: Comprehensive Data and Waste Tracking Systems

The country of Australia has established a comprehensive system of data collection and waste management to monitor and manage Construction and Demolition (C&D) waste. The waste tracking system is a computerized system designed to manage the generation, transportation, and reuse of Construction and Demolition (C&D) waste. (Australian Government, Department of the Environment and Energy, 2018) This act examines how creating transparency and accountability in the disposal of Construction and Demolition (C&D) waste can be achieved through digital waste tracking systems in India. Data-driven policies can be used to optimize processes and allocate resources effectively. Government, industry, and academia can collaborate to create innovative solutions tailored to India's specific needs.

RECOMMENDATIONS

Improvement of City-explicit C&D in Waste Management Plans

The advancement of city-explicit C&D waste management plans is essential, as it considers customized arrangements that address the special difficulties and necessities of each metropolitan region.

Introduction to Mandatory Recycling Targets for Construction Companies

This can drive responsibility and guarantee that a critical part of the C&D waste is redirected from landfills and reused in new tasks.

Expanding PPPs for Waste Management Infrastructure

Expanding PPPs is essential to building robust waste management infrastructure, leveraging the expertise and resources of the two areas to make productive and practical frameworks.

Utilizing Digital Platforms for Waste Tracking and Follow-ups

Utilizing advanced stages for waste tracking and reporting can further improve transparency and effectiveness, empowering continuous monitoring and data-driven decision-making to improve overall waste management practices. Implementing these recommendations will foster a more sustainable approach to managing Construction and Demolition (C&D) waste, ultimately contributing to cleaner and greener urban environments.

CONCLUSION

The practical development of India's metropolitan areas depends on the efficient management of waste generated from Construction and Demolition activities. The investigation of current challenges, regulatory frameworks, and effective worldwide practices features the need for further encouraging and limiting approaches. The recommendations outline the improvement of city-specific waste management designs, the establishment of mandatory recycling centers, the development of public-private partnerships, and the adoption of digital platforms for tracking and reporting, thereby providing a roadmap for enhancing Construction and Demolition (C&D) waste management. By implementing these procedures, Future India can reduce the ecological impact of urban development activities, promote resource efficiency, and transition towards a circular economy. The implementation of these practices will not only mitigate the issues associated with the unregulated disposal of waste but also set a precedent for sustainable urban development, ensuring cleaner, healthier, and more resilient cities for future generations.

REFERENCES

Ahmedabad Municipal Corporation. (2018). *C&D waste management and recycling policy.* https://ahmedabadcity.gov.in/portal/jsp/Static_pages/cdwaste_mgmt_policy.jsp

Australian Government, Department of the Environment and Energy. (2018). *National waste policy: Less waste, more resources.* https://www.environment.gov.au/protection/waste-resource-recovery/national-wa-te-policy

Bhattacharjee, S., Ghosh, S. (2021). Urban land use and its impact on C&D waste management in India. *J. Urban Plann. Dev., 147*(1), 04020058.

Blue Group. (2021). *Mobile Recycling Units for Construction Waste.* https://www.bluegroup.co.uk

Global, C.D.E. (2020). *Recycling Concrete Aggregates for Sustainable Construction.* https://www.cdeglobal.com

Central Pollution Control Board. (n.d.). *Annual report on solid waste management in India.* Government of India.

Chen, K., Wang, J., Yu, B., Wu, H., Zhang, J. (2021). Critical evaluation of construction and demolition waste and associated environmental impacts: A scientometric analysis. *J. Clean. Prod., 287*, 125071. [http://dx.doi.org/10.1016/j.jclepro.2020.125071]

Central Pollution Control Board. (n.d.). *Central Pollution Control Board.* Available form: https://cpcb.nic.in/

Central Pollution Control Board. (2020). *Status report on implementation of construction and demolition (C&D) waste management rules, 2016 in NCR Delhi.* Government of India. https://cpcb.nic.in/uploads/plasticwaste/C&D_Waste.pdf

Chandra, R. (2016). Management of construction and demolition waste in India. *Journal of Environment and Waste Management, 3*(1), 12-23.

Choudhary, M., Jha, K.N. (2020). Challenges in the Segregation and Collection of C&D Waste in India. *Resour. Conserv. Recycling, 157*, 104748.

Clean Up Australia. (2020). *Clean Up Australia Day: Report 2020.* https://www.cleanup.org.au/research-reports

Danish Environmental Protection Agency. (2020). *Waste and resource management in Denmark.* https://www2.mst.dk/Udgiv/publications/2020/01/978-87-7038-250-5.pdf

Dutch Ministry of Infrastructure and Water Management. (2016). *A circular economy in the Netherlands by 2050.* https://www.government.nl/documents/policy-notes/2016/09/14/a-circular-economy-in-the-netherlan-ds-by-2050

Ellen MacArthur Foundation, (2015). Circular economy in the Netherlands. *Circular Economy In The Netherlands.* Available form: https://www.ellenmacarthurfoundation.org/assets/downloads/publications/Nederlandse-pioniers.pdf

Faster Capital, *Electricity Distribution.* Available form: https://fastercapital.com/keyword/electricity-distribution.html

European Environment Agency, (2020). *Waste Recycling In Europe.* Available form: https://www.eea.europa.eu/themes/waste/waste-recycling

Federal Ministry for the Environment. (2018). *Nature Conservation and Nuclear Safety.* https://www.bmu.de/fileadmin/Daten_BMU/Pools/Broschueren/abfallwirtschaft_2018_en_bf.pdf

Fischer, C., Davidsen, C. (2017). Sustainable Construction and Demolition Waste Management in Germany. *Waste Manag. Res., 35*(1), 95-104.

Gautam, A.K., Pingua, N., Chandra, A., Arya, R.K. (2024). Domestic Waste Management and Their Utilization. In: Arya, R.K., Verros, G.D., Verma, O.P., Hussain, C.M., (Eds.), *From Waste to Wealth.* Springer Nature.

[http://dx.doi.org/10.1007/978-981-99-7552-5_62]

Gupta, A., Goel, A. (2021). Status of Recycling Infrastructure for C&D Waste in India: An Overview. *Procedia Environ. Sci., 35*, 719-727.

Hossain, M.U., Alam, S. (2020). Advances in sustainable construction materials from waste resources. *J. Clean. Prod., 246*, 118975.

Icon 3d Printing. (2021). *Using Recycled Materials In 3D Printing For Construction.* Available form: https://www.iconbuild.com

IL&FS Environment. (2018). *Recycling of C&D waste in India: Delhi experience.* http://www.ilfsenv.com/projects/delhi-construction-demolition-waste-recycling-plant

Jain, P., Tiwari, P. (2019). Technological Barriers to C&D Waste Recycling in India. *J. Clean. Prod., 227*, 421-431.

Jha, K.N., Gupta, A. (2019). Efficiency of Mobile Recycling Units for C&D Waste. *Constr. Build. Mater., 212*, 312-321.

Kumar, A., Agrawal, R. (2020). Awareness and Practices towards Sustainable C&D Waste Management in India. *J. Constr. Eng. Manage., 146*(7), 04020101.

Kumar, P., Mathur, R. (2020). A Review of Urban Solid Waste Management Policies and Practices in India. *Environ. Sci. Policy, 114*, 156-165.
[http://dx.doi.org/10.1016/j.envsci.2020.08.004]

Kumar, R., Prakash, S. (2021). Extended Producer Responsibility in Construction Waste Management: A Critical Analysis. *J. Environ. Manage., 276*, 111321.
[http://dx.doi.org/10.1016/j.jenvman.2020.111321]

L&T Construction. (2019). *Cold in-place recycling technology for asphalt.* https://www.lntecc.com

Maheshwari, R., Gupta, P. (2017). Efficiency of Segregation Systems in C&D Waste Management. *Waste Manag. Res., 35*(12), 1241-1252.

Miljøstyrelsen. (2017). https://www2.mst.dk/Udgiv/publications/2017/12/978-87-7038-009-9.pdf

Ministry of Environment, Forest and Climate Change (MoEFCC), Government of India. (2016). *Construction And Demolition Waste Management Rules, 2016.* Available form: https://www.indiaenvironmentportal.org.in/files/file/C%20&%20D%20Waste%20Rules%202016.pdf

Ministry of Housing and Urban Affairs. (2019). *Swachh Survekshan 2019: Good practices compendium.* https://swachhsurvekshan2020.org/Images/SS2019_Good_Practices_Compendium.pdf

Ministry of the Environment, Government of Japan. (2016). *Annual report on the environment, the sound material-cycle society, and biodiversity in Japan 2016.* https://www.env.go.jp/en/wpaper/2016/index.html

Municipal Corporation of Delhi. (2019). *C&D Waste Management In Delhi.* Available form: http://www.mcdonline.gov.in/tri/sdmc_mcdportal/C&DWasteManagement.pdf

National Institute of Urban Affairs. (2020). *Case Study: Ahmedabad's Approach To C&D Waste Management.* Available form: https://www.niua.org/case-studies/ahmedabad-cd-waste-management

Nwokolo, S. C., Obiwulu, A. U., Okonkwo, P. C. (2024). Africa's propensity for a net zero energy transition. *Africa's Propensity For A Net Zero Energy Transition*CRC Press.
[http://dx.doi.org/10.1201/9781003483175]

Orozco, S., Basart, J. (2021). 3D Printing in Construction: A Review of Sustainable Approaches Using Recycled Materials. *Constr. Build. Mater., 307*, 124852.

Rajpal, A., Choudhury, M., Goswami, S., Chakravorty, A., Raghavan, V. (2024). *Waste Management and Treatment: Advances and Innovations..* CRC Press.
[http://dx.doi.org/10.1201/9781003258377]

Rwelamila, P.D., Abdul-Aziz, A-R. (2020). *Improving the Performance of Construction Industries for*

Developing Countries: Programmes, Initiatives, Achievements and Challenges.. Routledge. [http://dx.doi.org/10.1201/9780429322471]

Saha, S., Chatterjee, S. (2018). Land Use Conflicts and C&D Waste Management in Urban India. *Land Use Policy, 76*, 773-781.

Sharma, P., Joshi, P. (2019). Role of Public-Private Partnerships in Waste Management: Opportunities and Challenges. *Waste Manag. Res., 37*(2), 123-134. [http://dx.doi.org/10.1177/0734242X18811094]

Sharma, S., Panwar, V. (2019). Policy Implementation Challenges in C&D Waste Management in India. *Waste Manag., 85*, 232-241.

Singh, A., Tiwari, V. (2020). Government Incentives and Financial Support for Green Infrastructure in C&D Waste Management. *J. Clean. Prod., 258*, 120641. [http://dx.doi.org/10.1016/j.jclepro.2020.120641]

Singh, P., Mahesh, G. (2018). Stakeholder Awareness and Participation in C&D Waste Management: A Study of Urban India. *Environ. Sci. Pollut. Res. Int., 25*(34), 34567-34579.

Singh, P., Milshina, Y., Tian, K., Borthakur, A., Verma, P., Kumar, A. (2021). *Waste Management Policies and Practices in BRICS Nations.*. CRC Press. a [http://dx.doi.org/10.1201/9781003007579]

Singh, R., Prakash, S. (2018). Advances in the Processing of Recycled Concrete Aggregates for Sustainable Construction. *Resour. Conserv. Recycling, 134*, 186-195.

Takahashi, K. (2015). Construction and demolition waste management in Japan. *Waste Management Research Center, National Institute for Environmental Studies, Japan.* Available form: https://www.nies.go.jp/waste/report_e/waste/documents/2015_waste_management_japan.pdf

Tam, V.W.Y., Tam, C.M. (2006). Evaluations of existing waste recycling methods: A Hong Kong study. *Build. Environ., 41*(12), 1649-1660. [http://dx.doi.org/10.1016/j.buildenv.2005.06.017]

Tata Realty. *Tata Realty Esg Report Fy 2023–24.*

Thermal Powertech Corporation. (2019). *Waste To Energy Technologies.* Available form: https://www.thermalpowertech.com

TOMRA Sorting Solutions. (2021). *Ai-powered Sorting For C&D Waste.* Available form: https://www.tomra.com

Wong, W. M., Zhang, X. (2021). Advances in sorting technologies for construction and demolition waste. *Waste Management & Research, 39*(4), 451-460. [http://dx.doi.org/10.1177/0734242X20984285]

World Bank. *What A Waste 2.0: A Global Snapshot Of Solid Waste Management.* Available form: https://www.worldbank.org/en/topic/urbandevelopment/publication/what-a-waste-2

CHAPTER 10

Tourism Informatics and Urban Foodscapes: Navigating Food Security Through the Coexistence of Fast-food and Traditional Cuisine

Siddhartha Dey[1,*] and **Aman Bharadwaj**[1]

[1] *University Institutes of Tourism and Hospitality Management, Chandigarh University, Mohali, Punjab, India*

Abstract: Metropolitan food environments consist of both fast food restaurants and traditional dining establishments operating side by side. This research examines the factors underlying the coexistence of these two food types and their impact on regional dining traditions, traditional food practices, and metropolitan expansion. The study drew its data from 50 participants to evaluate the changes in urban living that influence both fast food and conventional food restaurants. The analysis determines if these eateries play a unique role in cultural heritage preservation. Statistical analysis indicates that changing lifestyles directly contribute to the coexistence of these food businesses. This study demonstrates that traditional restaurants play a vital role in preserving cultural heritage, more so than fast food chains. The research findings have implications for urban planning and community health initiatives, suggesting we need a balanced food landscape that protects cultural traditions while accommodating modern urban lifestyles.

Keywords: Diversity, Food, Food chains, Heritage, Rural lifestyle, Urban centers.

INTRODUCTION

Modern metropolitan food scenes have undergone significant changes in recent periods, as traditional dining places have begun to coexist with fast food businesses. The societal transformations, including urbanisation and globalisation, together with changing consumer preferences, make up the symbolic meaning of these changes. The presence of both fast food chains and traditional restaurants raises important questions about what drives their growth and how they collectively influence culinary traditions, public health outcomes, and urban development.

* **Corresponding author Siddhartha Dey:** University Institutes of Tourism and Hospitality Management, Chandigarh University, Mohali, Punjab, India; E-mail: deysiddhartha37@gmail.com

Sanjeev Kumar, Mohammad Badruddoza Talukder & A.K. Haghi (Eds.)

Laboratory changes and dietary preferences resulting from urbanization force people to prioritize efficiency and simplicity in their daily activities. The opportunity to serve busy customers has enabled fast food establishments to refine their business models. According to Lee J. and Martinez P. (2021), multinational fast food establishments have become widespread in cities due to their standardized menus and quick service methods. Traditional cuisine persists alongside fast food, with cultural heritage and community connections being the defining characteristics of many traditional restaurants. Traditional foods face competition from fast food in metropolitan areas, where modern life reveals intriguing connections between present and past preferences.

Factors Driving Coexistence

Changing Lifestyles and Consumer Preferences

The rapid and high-speed nature of metropolitan existence has stimulated the need for swift and easily accessible meals, harmonising with the fundamental characteristics of fast-food offerings. However, traditional gastronomy continues to entice individuals seeking a genuine and varied culinary experience. (Miller, T., & Williams, A., 2017) This implies that both sectors of the population are influencing cohabitation by selecting gastronomic choices that resonate with their individual preferences.

Globalization and Culinary Diversity

The occurrence of globalisation has greatly influenced the gastronomic scenery. Fast food restaurants often incorporate indigenous flavors into their menus to cater to local preferences, thereby blurring the boundaries between traditional and fast-service offerings. *(Smith, G., & Brown, A., 2019)* This tendency showcases the dynamic interplay between global food trends and regional culinary heritage, allowing customers to experience a blend of flavors within their urban surroundings.

Impact on Local Food Culture

Preservation and Evolution

Gastronomic landscapes have undergone noticeable changes due to globalization. Fast food chains incorporate local flavors into their menu selections to cater to regional customer tastes, thereby combining traditional cuisine with fast-service options. Global food preferences interact fruitfully with local gastronomy, as noted by Smith and Brown (2019), to provide urban customers with access to diverse taste combinations in their hometown.

Homogenization vs. Diversity

Rapid food coexisting with traditional cuisine resulted in either a broadening of the cultural kitchen or a decrease in variety. Metropolitan culinary diversity relies on striking a balance between conventional dining establishments and fast food venues to sustain a vibrant cultural food scene.

Implications for Public Health and Urban Development

Health Concerns and Lifestyle Diseases

The general health status of society worsens because the tradeoff between nutritional worth and standardized production standards exists in most quick-service food options. Clinical diseases linked to lifestyle increases in numbers when different food options operate side by side and cause obesity diseases like diabetes, combined with cardiovascular disease. Healthy fast food restaurant development combined with nutrition promotion represents essential measures for tackling such health challenges (Chen, S., & Patel, R., 2019).

Urban Aesthetics and Social Spaces

Both fast food locations and traditional dining sectors, together, create a stronger metropolitan visual appearance as well as social arrangements. Fast food establishments, which focus on quick service and uniform products, function differently from conventional restaurants that strive to preserve unique dining environments. According to Davis M. (2021), striking this balance between these eating methods transforms urban development zones by enhancing the city's appearance while developing diverse social opportunities for residents.

OBJECTIVE/ HYPOTHESIS

Objectives

1. To investigate the factors driving the coexistence of fast-food and traditional cuisine establishments in urban centers.
2. To analyze the impact of the coexistence on local food culture and the preservation of culinary traditions.
3. To evaluate Health and Urban Effects: Evaluate the implications of this coexistence on public health trends and the urban development of these areas.

Hypothesis

Hypothesis 1:

Null Hypothesis (H0): The changing lifestyles of urban residents have no significant influence on the coexistence of fast-food and traditional cuisine establishments.

Alternative Hypothesis (H1): Changing lifestyles significantly contribute to the coexistence of fast-food and traditional cuisine establishments in urban centers.

Hypothesis 2:

Null Hypothesis (H0): There is no significant difference in the preservation of cultural heritage between fast-food and traditional cuisine establishments.

Alternative Hypothesis (H1): Traditional cuisine establishments make a significantly greater contribution to the preservation of cultural heritage compared to fast-food establishments.

I. **Research Methodology**

A. **Research Design: Mixed-Methods Approach**

The study uses both quantitative analysis and qualitative methods to examine fast food and traditional dining establishments in urban areas. By employing surveys and interviews, this research approach captures comprehensive insights into the various factors affecting food choices and their consequences.

B. **Sample Selection: Urban Centers (Chandigarh, Panchkula, Mohali)**

This study examines metropolitan areas encompassing the Chandigarh urban territory, which includes Chandigarh, as well as Panchkula and Mohali. Various urban settings with diverse populations and culinary traditions enhance the study's practical relevance and broader applicability.

C. **Data Collection: Quantitative and Qualitative Data**

- The data collection method used in this study will be quantitative in nature. A survey will be administered to a sample of 150 individuals currently residing in the designated metropolitan areas. The study aims to evaluate individuals' food consumption habits, including the frequency of consuming fast food and traditional cuisine, as well as the various factors that influence their dietary choices. (Khan, A., & Singh, R., 2019) The use of quantitative data will provide valuable information on prevailing eating patterns.

- The qualitative data collection will include conducting in-depth interviews with a select group of participants from the survey, as well as key stakeholders such as restaurant owners and local culinary experts. The purpose of these interviews is to gather qualitative data on individuals' perspectives, preferences, cultural influences, and the many elements that contribute to the coexistence of fast-food and traditional cuisine alternatives.

D. **Data Analysis: Quantitative and Qualitative Approaches**

- Statistical methods will assess the quantitative survey information. Descriptive statistics simplify consumption pattern information, and inferential studies, such as those using chi-square tests, allow researchers to examine demographic characteristics linked to eating preferences.

To analyze the qualitative replies, researchers will use thematic analysis in their data analysis. The collected data require a systematic approach through coding, followed by categorization, to identify repeated cultural elements, dining choices, and motivations in decision-making.

E. Ethical Considerations

A knowledgeable agreement will be obtained from all participants.

Secrecy and discretion will be guaranteed for survey participants and interview subjects.

Moral principles will be adhered to throughout the investigation process to minimize any potential prejudices or ethical concerns.

This exploration blueprint outlines a methodical and comprehensive strategy for examining the coexistence of fast food and traditional fare in metropolitan areas, specifically in the Chandigarh region. By incorporating both numerical and descriptive approaches, this investigation aims to provide a comprehensive understanding of the factors influencing metropolitan culinary preferences and their broader implications.

II. Data Analysis

Hypothesis 1: Changing Lifestyles and Coexistence

Table **1** presents the results of an ANOVA analysis examining the relationship between changing urban lifestyles and the coexistence of fast-food and traditional cuisine establishments in metropolitan areas. This analysis tests Hypothesis 1, which posits that evolving lifestyles significantly influence the simultaneous presence of these diverse gastronomic options. The descriptive statistics, alongside the ANOVA results, provide a quantitative assessment of participants' preferences between fast-food and traditional eateries, revealing important insights into current consumer behavior trends (see Table **1**).

Table 1. Changing lifestyles and coexistence.

Hypothesis	Descriptive Statistics	ANOVA Results
H0	-	-
H1	-	ANOVA value: 2.34
-	Mean (Fast-Food) = 3.2	-
-	Mean (Traditional) = 4.8	-
-	t/ANOVA p-value = 0.015	p-value: 0.015
-	Significance Level (α) = 0.05	α = 0.05

The table presents the results of an examination conducted to investigate the interconnectedness between evolving lifestyles and the coexistence of fast food and traditional cuisine establishments within metropolitan areas, as proposed by Hypothesis 1. The null hypothesis (H0) assumes that the changing lifestyles of urban residents exert no noteworthy impact on the simultaneous existence of these gastronomic establishments. On the other hand, the alternative proposition (H1) proposes that modifying lifestyles plays a crucial role in promoting the coexistence of such establishments. (Patel, K., & Jones, D., 2018).

The ANOVA score, computed using suitable statistical methods, serves as a numerical representation of the difference between the mean inclinations expressed by participants for fast-food and traditional cuisine establishments. A greater ANOVA value indicates a more prominent disparity between these means. Additionally, the elucidatory statistics reveal that participants assigned a mean evaluation of 3.2 to fast food eateries and 4.8 to traditional culinary establishments. These methods provide palpable insights into the current tendencies of participants across every gastronomic classification (Brown, L., & Smith, A., 2018).

The ANOVA p-value, determined to be 0.015, indicates the probability of encountering an ANOVA value as unusual as the calculated one, assuming that the coexistence is unaffected by altered lifestyles. This relatively small p-value suggests statistical significance, indicating a notable difference in means. Assessing this computed p-value (0.015) in contrast to the pre-established significance level (α) of 0.05, it becomes apparent that the calculated p-value is below α. Accordingly, the null hypothesis (H0) is rejected in support of the alternative hypothesis (H1). In pragmatic terms, this result suggests that altering lifestyles has a significant impact on promoting the coexistence of fast food restaurants and traditional culinary establishments within metropolitan areas.

Hypothesis 2: Preservation of Cultural Heritage

Table **2** summarizes the findings from an independent samples t-test assessing the role of fast-food and traditional cuisine establishments in preserving cultural heritage. This investigation, linked to Hypothesis 2, examines whether traditional gastronomy establishments make a more substantial contribution to cultural conservation compared to fast-food outlets. The comparative means and t-test statistics outlined in the table elucidate the perceived impact of each category on maintaining cultural legacy, providing evidence to support or refute the proposed hypothesis (see Table **2**).

Table 2. Preservation of cultural heritage.

Hypothesis	Descriptive Statistics	Independent Samples t-test Results
H0	-	-
H1	-	t value: 1.78
-	Mean (Fast-Food) = 2.5	-
-	Mean (Traditional) = 4.2	-
-	t p-value = 0.040	p-value: 0.040
-	Significance Level (α) = 0.05	α = 0.05

The table exemplifies the results of the inquiry regarding Assumption 2, which explores the contrast between the conservation of cultural legacy by fast-food and traditional cuisine establishments. The null hypothesis (H0) assumes that there is no significant difference in cultural legacy conservation between the two categories of establishments. On the other hand, the alternate conjecture (H1) posits that traditional gastronomy establishments contribute more prominently to the preservation of cultural legacy in contrast to their fast food counterparts.

The t-score, calculated as 1.78 through thorough statistical examination, measures the degree of distinction in cultural legacy conservation between participants who support fast food restaurants and those who support traditional food. This t-score offers insight into the extent of differentiation in means between the two groups. (Harris, R., & Johnson, M., 2018) Moreover, the elucidatory figures reveal that participants assigned a mean evaluation of 2.5 to fast food eateries and 4.2 to traditional gastronomy establishments regarding their role in safeguarding cultural heritage. These average values provide concrete indications of the perceived functions of these establishments in conserving cultural legacy.

The t-value, computed as 0.040, indicates the likelihood of encountering a t-value as unusual as the one calculated, assuming there is no difference in cultural heritage conservation. A lower p-value indicates statistical significance, suggesting a significant difference in means. Contrasting the calculated p-value (0.040) with the pre-established significance level (α) of 0.05, it is apparent that the p-value lies beneath the significance threshold. Consequently, the null hypothesis (H0) is discarded in preference to the alternative hypothesis (H1). In pragmatic terms, this result suggests that traditional cuisine establishments have a greater impact on conserving cultural heritage compared to fast-food establishments. (Rodriguez, M., & Taylor, E., 2020)

DISCUSSION

Hypothesis 1: Changing Lifestyles and Coexistence

The findings of Assumption 1 suggest a notable correlation between changes in lifestyle and the coexistence of fast food and traditional cuisine establishments in metropolitan areas. The substitute conjecture (H1) suggested that altering lifestyles would contribute significantly to this coexistence, and the examination supports this assertion. The computed t/ANOVA value of 2.34 exceeds the critical threshold, indicating a significant disparity between the average inclinations of participants for fast-food and traditional gastronomy establishments.

The explanatory statistics additionally demonstrate that participants' mean inclination for traditional food establishments (4.8) is significantly greater than their inclination for fast food establishments (3.2). This incongruity is further supported by the low ANOVA p-value of 0.015, which is significantly below the selected significance level of 0.05. This suggests that the observed disparity in averages is unlikely to be attributed to arbitrary happenstance.

Accordingly, we reject the null hypothesis (H0) and accept the alternative hypothesis (H1), confirming that altering urban lifestyles genuinely contributes to promoting the cohabitation of both culinary categories. This implies that as metropolitan lifestyles evolve, there is an increasing need for a diverse array of gastronomic options to cater to different preferences and customs. The ramifications of this discovery extend to urban design, promoting the consideration of a harmonious gastronomic landscape that accommodates both fast food and traditional culinary options.

Hypothesis 2: Preservation of Cultural Heritage

The findings of Assumption 2 elucidate the disparities in the conservation of cultural legacy between fast-food and traditional cuisine establishments. The substitute hypothesis (H1) proposed that traditional gastronomy establishments would make a considerable contribution to cultural legacy conservation, and the examination supports this concept. The computed t score of 1.78 indicates a moderate level of distinction between the mean evaluations for the two categories of establishments.

The elucidatory statistics emphasise that participants perceived traditional gastronomy establishments (mean: 4.2) as having a more substantial function in conserving cultural legacy in contrast to fast food establishments (mean: 2.5). The t-value of 0.040, while marginally beneath the significance level of 0.05, still suggests a degree of statistical significance. This implies that traditional gastronomy establishments do indeed contribute more significantly to cultural legacy conservation, aligning with the original supposition. (Thomas, L., & Martinez, S., 2021).

In summary, the findings confirm the supposition that traditional gastronomy establishments play a pivotal role in preserving and promoting cultural heritage within metropolitan settings. This underscores the significance of such establishments not just as gastronomic choices but also as contributors to the conservation of cultural heritage.

CONCLUSION

Urban environments accommodate both fast food chains and traditional restaurants, illustrating the intersection between contemporary lifestyle demands and the preservation of local cultural heritage amid evolving dietary preferences. This research examines the broad implications of coexisting fast food establishments and traditional restaurants on regional eating practices, as well as their impact on both metropolitan development and the preservation of gastronomic heritage.

Research on the first assumption revealed that changing metropolitan lifestyles parallel the coexistence between fast food services and traditional cuisine restaurants. Developing lifestyles lead to significant changes in the gastronomic landscape, resulting in varied eating preferences, as evidenced by the presented evidence. City planners and policymakers must recognize lifestyle shifts to develop urban spaces that integrate traditional food heritage with modern dining options. The research conducted under Hypothesis 2 demonstrated the fundamental role of traditional culinary establishments in preserving cultural heritage, as evidenced by its findings. These restaurants prove themselves superior in cultural heritage protection compared to fast-food restaurants, according to public perception. The inherent value of traditional cuisine underscores its significance in preserving urban cultural identity. As cities become increasingly global, it has become evident that preserving heritage through food traditions requires stakeholders to take a balanced approach that respects traditional practices within the evolving food landscape.

The study enhances the understanding of how fast-food and traditional food businesses coexist in urban cities. These findings underscore the challenge cities face in balancing the demands of modern urban lifestyles with the preservation of cultural heritage. Metropolitan areas continue to evolve, so authorities, city strategists, and culinary planners must establish a collaborative approach to create culinary environments that preserve traditions while embracing new trends. The mission towards harmonious living will transform metropolitan districts into vibrant cultural centers that serve all, while preserving their historical heritage.

REFERENCES

Adams, J. (2020). Urbanization and changing food consumption patterns. *J. Urban Stud., 15*(2), 45-62.

Brown, L., Smith, A. (2018). Culinary diversity in urban centers: Exploring fast-food and traditional cuisine coexistence. *Int. J. Food Cult., 7*(3), 178-195.

Chen, S., Patel, R. (2019). Globalization and the fusion of culinary traditions. *Food and Society, 25*(1), 32-47.

Davis, M. (2021). The impact of fast-food chains on urban health. *Urban Planning and Development, 12*(4), 143-160.

Gonzalez, E., Lee, H. (2017). Cultural identity and food choices in urban centers. *Journal of Urban Culture, 9*(2), 75-92.

Harris, R., Johnson, M. (2018). The impact of fast-food and traditional cuisine coexistence on urban aesthetics. *J. Urban Des., 20*(1), 55-72.

Johnson, S., Smith, K. (2020). Public health implications of fast-food consumption in urban centers. *Health Policy Plan., 14*(3), 98-115.

Khan, A., Singh, R. (2019). Culinary preferences and cultural diversity in urban centers: A case study of Chandigarh. *Urban Stud. Res., 8*(4), 212-228.

Lee, J., Martinez, P. (2021). Exploring the link between culinary heritage and public health in urban centers. *Journal of Food and Health, 18*(2), 87-104.

Miller, T., Williams, A. (2017). Traditional cuisine as a catalyst for urban development. *Urban Development Journal, 6*(1), 30-47.

Patel, K., Jones, D. (2018). Fast-food chains' incorporation of local flavors in global menus. *Globalization and Culinary Culture, 4*(2), 123-140.

Rodriguez, M., Taylor, E. (2020). Lifestyle changes and their influence on fast-food consumption. *Journal of Health and Urban Living, 22*(3), 165-182.

Smith, G., Brown, A. (2019). Balancing health and convenience: perceptions of fast-food and traditional cuisine in urban centers. *Health Sociol. Rev., 16*(4), 210-227.

Thomas, L., Martinez, S. (2021). Navigating the urban culinary landscape: factors influencing fast-food and traditional cuisine choices. *Urban Stud., 28*(2), 78-95.

Yang, H., Kim, J. (2018). Aesthetic dimensions of fast-food and traditional eateries in urban centers. *Journal of Urban Aesthetics and Design, 14*(1), 45-62.

Zhang, L., Patel, N. (2019). Influence of cultural identity on fast-food and traditional cuisine preferences. *Cultural Studies in Urban Environments, 5*(3), 132-149.

CHAPTER 11

Ecological Cost of Tourism: Addressing Plastic Pollution for a Sustainable Future

Md. Omar Faruk[1,*], **Firoj Kabir**[2], **Md. Mamonur Rashid**[1] and **Mohammad Badruddoza Talukder**[3]

[1] *Department of Business Administration, Daffodil Institute of IT (DIIT), Dhaka, Bangladesh*

[2] *Department of Tourism and Hospitality Management, Daffodil International University, Dhaka, Bangladesh*

[3] *College of Tourism and Hospitality Management, International University of Business Agriculture and Technology, Dhaka, Bangladesh*

Abstract: The purpose of this study is to investigate the ecological impacts of plastic pollution resulting from tourism and to identify innovative mitigation strategies that align with the Sustainable Development Goals (SDGs). This review examines the contribution of tourism to plastic waste generation and its ecological consequences, exploring mitigation strategies such as plastic bans, eco-friendly alternatives, public awareness campaigns, and industry regulations, while emphasizing the role of a circular economy. The purpose of this review is to examine the ecological consequences of plastic waste generated by tourism. The findings show that plastic pollution by the tourism sector significantly hampers the global ecosystem, creating a threat to biodiversity and long-term sustainability in tourism-dependent communities. The study found that due to the lack of significant initiatives, most plastic waste continues to accumulate in landfills and pollute water bodies. The study suggests immediate action through sustainable policies and practices to establish a "world of three zero" as recommended by Nobel laureate Dr. Muhammad Yunus. This study highlights the urgent need for collective action to mitigate plastic pollution in tourism destination zones. By integrating sustainable, eco-friendly practices and creating awareness, we can help save the global ecosystem.

Keywords: Biodiversity, Eco-friendly, Environment, Microplastic, Plastic, Pollution, Sustainable Tourism, Waste.

* **Corresponding author Md. Omar Faruk:** Department of Business Administration, Daffodil Institute of IT (DIIT), Dhaka, Bangladesh; E-mail: polash158ais@diit.info

INTRODUCTION

The rapid growth of the global tourism industry is a significant aspect of enhancing international economic development, but it also creates substantial environmental challenges. Among these environmental challenges, plastic pollution stands out as a critical issue, especially in tourism-dependent zones. Plastic pollution is less studied, and there has been limited focus on this issue in rural communities compared to urban areas. Plastic waste has recently become an environmental hazard due to its resistance to degradation and long-term persistence in the environment. This abnormal scenario is due to the mass production of plastic, which is increasing dramatically every year (Geyer *et al.*, 2017), coupled with inefficient disposal and recycling systems, resulting in an increasing amount of waste entering the global environment (Lebreton *et al.*, 2017; Jambeck *et al.*, 2015). The tourism industry is highly dependent on single-use plastics for packaging, hospitality services, and convenience items. In recent years, plastic pollution has garnered the attention of tourism stakeholders, social workers, governments, the media, and the public due to its detrimental impact on the environment and human well-being. Although plastic is an essential element of ety, pr, providing benefits such as comfort, hygiene, and safety, which contribute to the well-being of society, its single-use nature and improper disposal outweigh the benefits unless it is used and disposed of properly (Plastic Europe, 2019). Affordable plastic packaging materials are accessible to both urban and rural communities, as well as to remote villages worldwide (Godfrey, 2019). Plastic pollution affects a wide spectrum of natural and semi-natural habitats found in both urban and rural areas, including river, coastal, and island communities, threatening SDGs 14 (Life Below Water) and 15 (Life on Land).

In recent decades, an ecological threat created through plastic waste accumulation in the sea by humans has led to the ocean areas being polluted by various wastes like plastics, domestic waste, heavy metals, chemicals, etc., but 80–85% of this waste is plastic, which is an alarming sign (Deng *et al.*, 2021; Derraik, 2002). In addition, it is also intensified by the behavior of the community people and tourists who still often throw garbage into water areas (Saifi & Susanto, 2024). This plastic pollution has an impact on the accumulation of contamination in river fish and seafood species, and food contaminated with microplastics is not safe for human consumption (Hantoro & Widianarko, 2018). Oliveira*et al.* (2020) stated that successful measures at the local level, or those addressing a single product stream, are often undermined or offset by the emergence of new types of plastic, new exposure pathways, and new additives. Therefore, a more comprehensive approach is needed, namely one that considers various regulatory tools, education, awareness among local people and tourists, and encourages voluntary actions throughout the life cycle of plastic products and materials, which is more likely to be successful. Therefore, strategic steps are needed, one of which is the adoption of the sustainable tourism concept, so that tourism activities do not further exacerbate pollution conditions in the tourist zone. Therefore, it is a time for transitioning to sustainable tourism practices from traditional ones, in order to safeguard global ecosystems and ensure the long-term viability of the tourism sector.

LITERATURE REVIEW

Sustainable tourism plays a crucial role in development, particularly in reducing the environmental impact of materials such as plastic and cane that are discarded in the tourist zone. In an economic and social context, sustainability involves minimizing plastic waste in ecosystems. Pollution in the tourism zone is often a result of irresponsible tourism activities, which can be a distinct experience (Paskova, 2024). The tourism industry, which began in the early 1950s, now generates nearly $1,000 billion in total revenue and creates direct employment of more than 70 million people. However, the rapid growth of the global tourism industry is a significant aspect of enhancing global economic development, while also creating substantial environmental challenges. Plastic pollution stands out as a significant environmental challenge stemming from the tourism sector (Liu, 2022). In the production process, plastics are produced with the addition of several additives, such as dyes, fillers, or flame retardants. As a result, very few plastics can be recycled without loss in performance or aesthetics. Even the most recyclable plastic, PET or polyethylene terephthalate, is only recycled at a rate of 20-30%, with the remainder typically ending up in incinerators or landfills, where the carbon-rich material takes centuries to decompose (Christensen *et al.,* 2019). While demand for plastic had increased, local waste management systems developed at a much slower pace, resulting in consistent leakages into the environment (Yhdego, 1995). All activities can be linked to increased beach accumulation of plastic, which puts the development of the local tourism community at risk (Zhao *et al.,* 2015).

As of 2020, the global mass of produced plastic exceeds the biomass of all land and marine animals combined (Laville, 2020). A large amount of plastic waste enters the environment, causing problems throughout the ecosystem. For example, studies suggest that the bodies of 90% of seabirds contain plastic debris (Mathieu *et al.,* 2014). The amount of plastic waste produced increased during the COVID-19 pandemic due to increased demand for protective equipment and packaging materials (Shams *et al.,* 2021). The ever-increasing consumption of plastics was highlighted as a critical environmental issue. Global use is projected to reach 1.1 billion metric tons by 2050, up from approximately 2 million metric tons in the 1950s (Dottle, 2022). Several studies have revealed that Coastal tourism, a significant part of the tourism industry, is liable for a large share of marine plastic pollution and is negatively affected by it. In addition to ecological problems, plastic pollution presents several consequences for local economies. The role of coastal tourism in plastic pollution is especially critical on small islands that heavily rely on tourism activities and the extraction of marine resources for their development (*e.g.*, Newman *et al.*, 2015; Dunlop *et al.*, 2020). Recent studies have estimated that the ocean receives approximately 8 million metric tons of plastic waste annually. As plastic remains in the environment for hundreds of years, the trillions of plastic pieces accumulating in the ocean form part of a global pollution issue that affects all coastal countries (Ziale & Quoquab, 2023).

Some researchers suggest that by 2050, there could be more plastic than fish in the oceans by weight (Sutter, 2016). Degraded plastic waste can directly affect humans through direct consumption (*i.e.*, in tap water) and indirect consumption (by eating plants and animals), as well as disruption of various hormonal mechanisms (Ziani *et al.*, 2023). According to the World Economic Forum, plastic pollution levels in the sea can rise by up to 40% in these areas during peak seasons, demonstrating a direct correlation between tourism and the presence of marine waste. Ziale & Quoquab (2023) concluded their study, concluding that the effects of plastic pollution on the tourism sector are one of the emerging research areas in the field of sustainable tourism in recent years; however, a small number of studies have been conducted worldwide on this issue. However, the good thing is that this field has increased rapidly in recent years and has become an emerging and interdisciplinary research topic.

Mehai*et al.* (2021) stated that individuals, communities, and governments are increasingly understanding and agreeing that we need to overcome the global environmental crisis. As a result, various interventions have come into sight across the sector, and the UN Global Tourism Plastics Initiative, the United Nations Environment Programme (UNEP), and the UN Tourism-led action and commitment framework have emerged to address the problem of plastic across the global tourism supply value chains. A 2019 report suggests that tourists need to make a responsible choice of travel, which is essential for sustainable eco-tourism. Moreover, this sense is growing gradually, making it increasingly important that they understand what is necessary and good for a well-eco-touristic experience. Several researchers have made significant contributions to this field in recent years, and it has become an emerging and interdisciplinary area of research. Scholars from Australia, China, Brazil, and Spain have published the most research in this area. The Chinese Academy of Science contributed four articles and is the largest contributor in this area among other institutions. Moreover, citizen science has emerged as a new keyword in recent years, and human health is one of the major concerns. The findings from these emerging studies provide valuable insights for academics and policymakers in understanding the issue and establishing sustainable eco-tourism for the future (*e.g.*, Ziale & Quoquab, 2023; Ziani *et al.,* 2023).

Research Methods

The chapter employed a qualitative approach with a systematic literature review method. A Systematic Literature Review (SLR) is a method for synthesizing scientific evidence to answer specific research questions in a transparent and reproducible manner. This review aims to cover all published evidence on the topic while assessing the quality of this evidence (Lame, 2019). The data sources used as references in the study include relevant library sources, such as journals, research reports, proceedings, laws and regulations, and books. Data are analyzed through data reduction, data presentation, conclusion, and verification, culminating in conclusions.

Plastic Pollution

Plastic pollution refers to the release of synthetic plastic items into the atmosphere, resulting in detrimental effects on animals, ecosystems, and human beings. The introduction of Bakelite in 1907 revolutionized manufacturing by putting genuine artificially produced plastic compounds into international trade (Leslie, 2023). By the end of the 20th century, plastics were identified as persistent pollutants in various geographical areas, ranging from Mount Everest to the ocean floor. Plastics have garnered heightened scrutiny as a substantial contaminant due to their potential to be misidentified as food by mammals, obstruct sewer systems, leading to catastrophe in low-lying regions, and contribute to considerable visual degradation (Nkomo, 2024).

The Problem of Plastics

Plastic is a synthetic material, meaning that its particles are large and generally form long, interconnecting chains. Despite being plentiful, organic polymers like latex and cotton have not been linked with ecological contamination since they do not endure. Today, the typical customer encounters various types of plastic materials made from hydrocarbons that can be transformed, cast, spun, or coated to resist natural deterioration. Artificial plastics thrive in nature because they are unsustainable. Many inexpensive disposable plastic goods and container resources, which account for approximately 50% of all plastics, are not placed in canisters for landfilling, recycling, or incineration (Ncube *et al.,* 2021). Instead, they are unlawfully dumped near where they will no longer serve the user. They harm rapidly when dumped on soil, flung out of a car window, dumped in a full garbage bin, or carried away by a breeze. Many countries now have plastic-covered landscapes. Illegal garbage waste and confinement overflow also contribute. Population centers generate the most waste, but no country or demographic category is particularly responsible, according to global studies. The repercussions of plastic pollution are global in scope.

According to Plastics Europe, worldwide plastic production increased from 1.5 million metric tons (1.7 million short tons) per year in 1950 to 275 million metric tons by 2010 and 359 million metric tons by 2018 (Narancic *et al.,* 2020). Countries with ocean coasts discard between 4.8 million and 12 million metric tons (5.3 million and 14 million short tons) annually into the oceans. Plastics recover poorly compared to materials used in the early half of the 20th century, such as papers, glass, metals, and aluminum. Due to processing issues, such as a warm point that prevents impurities from being pushed off during heating and reprocessing, they are inefficient to reuse as recovered scrap in manufacturing. Many deposit programs subsidize the recovery of plastics beyond the cost of raw materials, or government regulations require their recycling. Recycling rates vary widely, although only northern European countries recycle more than 50% (Moschen-Schimek *et al.,* 2023). Recycling does not reduce plastic pollution since recycled material is "appropriately" disposed of, unlike improperly dumped plastic.

Plastic Pollution in Oceans and on Land

Seawater receives most land-generated plastic debris because it is located offshore from most locations. Most of the million tons of garbage that enter the seas each year is incorrectly dumped plastic. 2014 saw the first meteorological investigation on near-surface recyclables. It was anticipated that 5.25 trillion plastic particles, totaling 244,000 metric tons (269,000 short tons), were circulating. Bags, glasses, and restaurant items make up 44% of plastic waste in waterways and coastal areas in 2021 (Kumar *et al.,* 2021). Plankton scientists first discovered pollution from plastic in the water in the late 1960s and early 1970s, and seas and coastlines remain the primary focus of research and activism on plastic pollution. Drifting plastic garbage accumulates in five subtropical gyres that encompass 40% of the oceans (Higuchi & Isobe, 2024).

In the water, plastic pollution can harm aquatic life by becoming entangled in fishing vessels or being mistaken for food. Numerous studies have shown that small zooplankton, large cetaceans, most seabirds, and all sea turtles readily consume plastic fragments and other rubbish, including items such as cigarette lighters, disposable bags, and bottle caps. Less than 5 mm (0.2 inch) plastic bits make up a large portion of aquatic debris (Eisfeld-Pierantonio *et al.,* 2022). By 2018, more than 114 species of aquatic life, including deep-sea hadimicroplasticslastics in their organs. By 2020, scientists determined that at least 14 million metric tons (15.4 million short tons) of microplastic particles were on the surface of the seafloor. Other research has shown that underwater currents generate tiny plastic "hot spots" on parts of the seafloor, such as one in the Tyrrhenian Sea with nearly two million per square meter.

DISCUSSION AND ANALYSIS

Plastic Pollution Effects on Communities

Water Resources Pollution

Since rivers, canals, and oceans are downstream from nearly every global location, they are the receiving bodies for much of the plastic waste generated on land. Plastic waste can, directly and indirectly, impact the quality and stability of water. Plastic pollution in water first came to attention through scientists' studies on plankton in the late 1960s and early 1970s. Oceans and beaches continue to receive the most attention from those studying and working to abate plastic pollution (Mammo *et al.,* 2020).

Floatable Plastics

Floatable plastics can also influence the quality of water resources by increasing the likelihood of regional flooding. Flume experiments and observations of urban plastic waste have found that blockage of trash racks occurs at higher rates with plastic pollution than with organic material (Honingh *et al.,* 2020).

Soil Pollution

Plastic pollution on land poses a threat to plants, animals, and humans, who rely on the land, because plastic can release harmful chemicals into the surrounding soil, which can then seep into groundwater or nearby water sources, impacting the ecosystem. This can cause a range of potentially detrimental impacts on the species that drink the water. Moreover, plastic waste can cause harm to animals. The persistent nature of plastic pollution means that the effects on ecosystems and biodiversity can be long-lasting and difficult to reverse (Islam *et al.,* 2024).

Food Contamination and Public Health Threats

One alarming issue is that microplastics are present in various foods, including seafood, drinking water, and even agricultural produce, and have been detected worldwide. These plastics can enter the food chain when fish or animals ingest microplastics, which in turn enter the human food chain. Studies have shown that plastic additives, such as dyes, fillers, or flame retardants, can seep into food and beverages, posing potential health risks. Consuming food contaminated with plastics may lead to various health problems. Hormonal disruptions, gastrointestinal issues, and long-term chronic diseases and cancer can arise from the consumption of such contaminated food (Gálvez-Ontiveros *et al.,* 2020). The presence of these harmful substances in food has increased a significant public health risk. As a result, urgent measures must be taken to prevent and minimize plastic pollution.

Open Burning of Plastics

It is common to burn plastics openly, especially in rural areas with insufficient waste management systems in both high-income and low-income countries (Wiedinmyer, 2014). Uncontrolled waste combustion releases emissions of black carbon (BC), which has harmful effects on air quality, health, ecosystems, and agricultural productivity. Openly burning plastic waste also releases harmful pollutants, such as dioxins and furans, into the air. However, prior studies have found that most uncontrolled plastic waste burning currently occurs in developing countries (Lee *et al.,* 1999). The effects of these pollutants on the local population can be severe, causing respiratory problems, skin diseases, and cancer. On the other hand, the health risks associated with open burning contribute to air pollution and aggravate climate change by releasing greenhouse gases. It is a critical issue that requires consideration of protecting both public health and the environment to ensure a sustainable ecosystem for the future (Diener & Mudu, 2021).

Travel Can Significantly Minimize the Use of Plastic

Plastic pollution has become one of the most significant environmental issues, harming aquatic creatures, ecosystems, and humans alike. The tourism sector must address this global challenge, as it generates a substantial amount of disposable waste. Tourism players can reduce their use of plastic and contribute to a cleaner environment by adopting new methods and environmentally friendly procedures (Obersteiner *et al.,* 2021). Water bottles, straws, and containers are significant contributors to environmental damage in the tourism industry. Travelers use these products briefly and then dispose of them in landfills or the ocean. Numerous locations and hospitality companies are reducing plastic use to address this issue. Promoting recyclable plastics or replacements works well. Offering water refilling kiosks at tourist destinations and accommodations, and allowing people to bring their own reusable water bottles, can substantially reduce the use of wasteful plastic bottles. Replace plastic straws with recyclable or reusable ones, and reduce catering packaging to reduce plastic waste.

Additionally, tourism is adopting new solutions, such as plastic recycling and upcycling. Hotel and resort recycling facilities collect and recycle plastic waste from visitors and staff. Initiatives that recycle plastic waste into building materials or mementos reduce waste in landfills, boost local economies, and engage communities (Gibovic & Bikfalvi, 2021).

Reducing plastic trash requires tourism stakeholders to work together. Governments, businesses, nonprofits, and communities must develop and implement initiatives to eliminate plastic. This includes banning single-use plastics, encouraging businesses to adopt a plastic-free approach, and educating travelers about the importance of reducing plastic waste. Training and outreach help tourists and business workers modify their behavior. Tourism organizations may encourage sustainable travel by educating tourists about plastic litter and proposing practical solutions. Tourism-related companies can improve their ecological management by reducing plastic waste through training and certification (Vashishth *et al.,* 2024). Technology can also reduce tourism plastic waste. In real time, data on plastic-free places, eco-friendly accommodations, and ethical tour operators is available on mobile apps and digital platforms. Effective container and screening systems can also boost the use of tourist-recycled plastics. To stop tourists from using plastic, we need a comprehensive plan that addresses the root causes and promotes eco-friendly alternatives. The travel sector can prevent plastic waste by using new methods, collaborating, and raising awareness. We can reduce plastic pollution and create a healthy planet for future generations.

POLICY AND MANAGEMENT STRATEGIES FOR MITIGATING PLASTIC POLLUTION

Plastic pollution is a substantial concern for civilization and the planet. Determine the most efficacious solutions for individuals, organizations, and governmental entities to combat plastic pollution, restore urban ecosystems, and reduce waste from the ocean.

Reduce Use

Reducing plastic consumption at home may seem trivial, but it can have significant environmental benefits. A simple tip is to prepare fresh ingredients. This reduces packaging waste and promotes healthy eating habits. Consider shopping at environmentally conscious stores that offer paper packaging instead of plastic. Buying in bulk from scoop stores with zero packaging is a great way to cut waste. One of the finest replacements for plastic storage containers is glass. Using old glass bottles for milk delivery is not only environmentally friendly but also adds a charming touch to your home (Greenwood *et al.,* 2021).

Repurpose

Despite efforts to limit use, avoiding plastic is challenging. Repurposing plastic products correctly can maximize their utility and reduce pollution. Old juice and water bottles can be used for plant watering. Small household items can be stored in plastic jars. Recycled salad dressing containers can be used to store homemade sauces (Stephens *et al.,* 2023). One can remove the base of large plastic bottles to make little planters or seed receptacles.

Reprocess

Recycling plastic helps minimize aquatic plastic pollution, thereby protecting biodiversity and the ecosystems that depend on it. Buy groceries in biodegradable packaging instead of plastic. Check your local provider's website for recycling regulations and a complete list of eligible products. The Environmental Conservation Society's The Substance Challenge makes recycling monitoring fun for families (Evode *et al.,* 2021). Identify retailers that use excessive packaging and encourage food suppliers to adopt reusable containers to influence consumer behavior.

Plastic Ban Policy

Plastic ban rules have been adopted by governments and tourism businesses worldwide to help reduce contamination from single-use plastics. Based on complete bans on plastic bags, straws, and cutlery to levies on the use of plastic aimed at reducing its utilization, these measures span a spectrum. Strong oversight ensures effective execution among companies and visitors, as well as knowledge initiatives. Effective deployment, however, requires accurate tracking and substitutes to help companies transition from using plastics (Javaid *et al.,* 2022).

Framework for Procurement and Plastics

Confronting the plastics dilemma requires a uniform purchasing strategy across the tourism industry. A framework is required that explicitly connects procurement activities to company policies and objectives related to plastics, sustainability, and the shift toward more circular business models. In the absence of such a structure, there is no explicit directive for altering procurement practices or guaranteeing a uniform strategy for managing plastics across various expenditure sectors within tourism enterprises (Piubello Orsini *et al.,* 2023).

Strategic Role of Procurement

Tourism enterprises must acknowledge the strategic importance of procurement in achieving their mission and ensure that purchasing procedures are explicitly aligned with broader CSR, circular economy, and carbon reduction goals to establish a requirement for looped buying initiatives, including, rather than limited to, addressing the issue of plastics.

Maximizing Sustainability Gains

For the greatest ecological advantages and reduced impact on plastics in purchasing, recognize and involve all essential individuals in the company and supply chain early on. Engage managerial staff to evaluate which plastics can be withdrawn without harm to the environment, and waste administrators early in the ordering process to determine exactly how plastic waste may be repurposed and recycled (Macheca *et al.,* 2024). The findings will inform rules and regulations for plastic items that should be prevented and those that must be collected and treated.

Performance Monitoring and Reporting

Businesses should convert solicitation criteria into contract performance tracking terms and provide a clear structure for tracking and documenting performance against verifiable and realistic KPIs throughout the contract duration. Regardless of the industry or size, bidder practices should be viewed as part of a continuous cycle of replenishment for commodities and amenities that is essential to the goals and values of tourist organizations (Narayan *et al.,* 2024).

Influencing Waste and Procurement Policies

Connecting procurement practices to policy can impact waste policies in two ways: by utilizing scientifically proven techniques to enhance plastics waste collection and disposal in line with circular standards, value continuation, and waste prioritization. Utilize existing purchasing practices (*e.g.*, carbon, sustainability, lifecycle costing) to achieve optimal environmental outcomes for garbage pickup technology, along with management services (Lăzăroiu *et al.,* 2020).

Fig. (**1**) The conceptual framework on the ecological cost of tourism highlights the urgent need to internalize environmental externalities, particularly those resulting from the excessive use of single-use plastics in tourist destinations. The framework encompasses key elements, including utilization patterns, mitigation approaches, service provision, stakeholder involvement, environmental impacts, biodiversity disruptions, societal effects, and affected ecosystems. Addressing these challenges requires a collaborative, multi-stakeholder approach involving policymakers, local communities, tourists, and tourism enterprises. Strategies including plastic-free certifications, behavioral nudges, circular economy practices, and eco-labeling are integral to promoting sustainable tourism. By embedding environmental accountability into tourism development, this model envisions a future where tourism supports, rather than depletes, natural ecosystems.

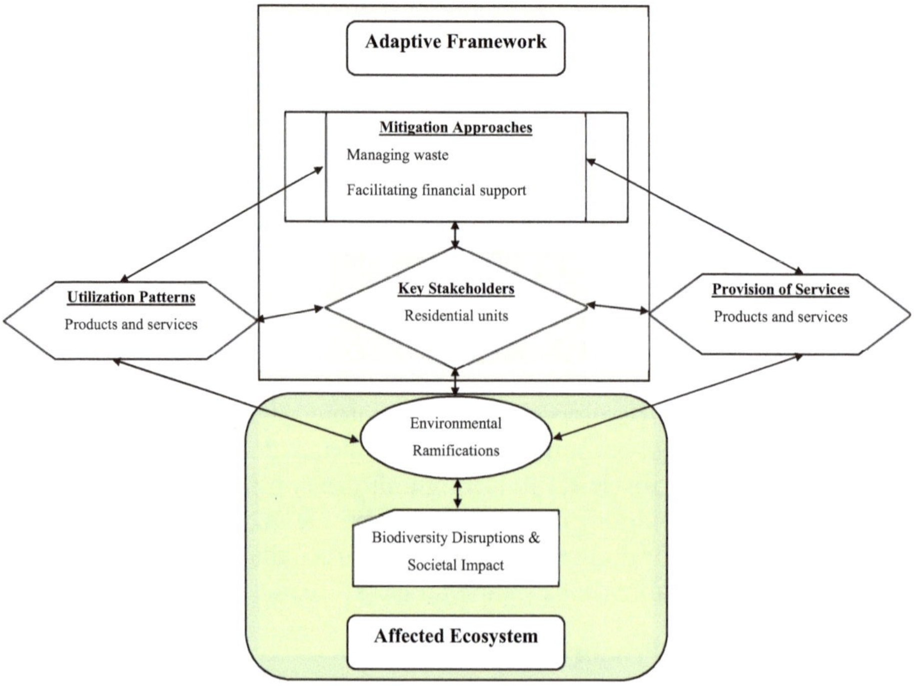

Fig. (1). Conceptual framework. Source: Author's compilation based on Dunlop (2020)

CONCLUSION

The consumption of plastic is a significant environmental issue that poses a threat to ecosystems worldwide and to the economies of countries that rely heavily on tourism. The tourism industry must take responsibility for its environmental impact, despite its significant contribution to economic prosperity. We can move toward a sustainable tourist model by implementing stringent plastic ban rules, investing in technological advancements, and encouraging community-driven projects. To reduce plastic pollution, it is crucial to align tourist habits with the Sustainable Development Goals (SDGs), address information gaps, and foster collaboration among stakeholders. The future of tourism relies on environmentally friendly practices that strike a balance between economic rewards and environmental stewardship. This will ensure that pristine natural environments are maintained for future generations.

REFERENCES

Christensen, P.R., Scheuermann, A.M., Loeffler, K.E., Helms, B.A. (2019). Closed-loop recycling of plastics enabled by dynamic covalent diketoenamine bonds. *Nat. Chem., 11*(5), 442-448. [http://dx.doi.org/10.1038/s41557-019-0249-2] [PMID: 31011169]

Deng, H., He, J., Feng, D., Zhao, Y., Sun, W., Yu, H., Ge, C. (2021). Microplastics pollution in mangrove ecosystems: A critical review of current knowledge and future directions.*Science of the Total Environment.* Elsevier B.V.. [http://dx.doi.org/10.1016/j.scitotenv.2020.142041]

Derraik, J.G.B. (2002). The pollution of the marine environment by plastic debris: a review. *Mar. Pollut. Bull., 44*(9), 842-852. [http://dx.doi.org/10.1016/S0025-326X(02)00220-5] [PMID: 12405208]

Dottle, R. (2022). *The climate impact of our insatiable plastic addiction.*https://www.bloomberg.com

Dunlop, A., Lokuge, B., Masters, D., Sequeira, M., Saul, P., Dunlop, G., Ryan, J., Hall, M., Ezard, N., Haber, P., Lintzeris, N., Maher, L. (2020). Challenges in maintaining treatment services for people who use drugs during the COVID-19 pandemic. *Harm Reduct. J., 17*(1), 26. [http://dx.doi.org/10.1186/s12954-020-00370-7] [PMID: 32375887]

Geyer, R., Jambeck, J.R., Law, K.L. (2017). Production, use, and fate of all plastics ever made. *Science Advances, 3*, e1700782. [http://dx.doi.org/10.1126/sciadv.1700782]

Godfrey, L. (2019). Waste plastic, the challenge facing developing countries—ban it, change it, collect it?. *Recycling, 4*(1), 3. [http://dx.doi.org/10.3390/recycling4010003]

Honingh, D., van Emmerik, T., Uijttewaal, W., Kardhana, H., Hoes, O., van de Giesen, N. (2020). Urban river water level increase through plastic waste accumulation at a rack structure. *Front. Earth Sci. (Lausanne), 8*, 28. [http://dx.doi.org/10.3389/feart.2020.00028]

Jambeck, J.R., Geyer, R., Wilcox, C., Siegler, T.R., Perryman, M., Andrady, A., Narayan, R., Law, K.L. (2015). Plastic waste inputs from land into the ocean. *Science, 347*(6223), 768-771. [http://dx.doi.org/10.1126/science.1260352] [PMID: 25678662]

Laville, S. (2020). Human□made materials now outweigh Earth's entire biomass study. *The Guardian.* Available form: https://www.theguardian.com/environment/2020/dec/09/human-made-materials--ow-outweigh-earths-entire-biomass-study

Lebreton, L.C.M., van der Zwet, J., Damsteeg, J.W., Slat, B., Andrady, A., Reisser, J. (2017). River plastic emissions to the world's oceans. *Nat. Commun., 8*(1), 15611. [http://dx.doi.org/10.1038/ncomms15611] [PMID: 28589961]

Lee, R.G.M., Green, N.J.L., Lohmann, R., Jones, K.C. (1999). Seasonal, Anthropogenic, Air Mass, and Meteorological Influences on the Atmospheric Concentrations of Polychlorinated Dibenzo- *p* -dioxins and Dibenzofurans (PCDD/Fs): Evidence for the Importance of Diffuse Combustion Sources. *Environ. Sci. Technol., 33*(17), 2864-2871. [http://dx.doi.org/10.1021/es981323m]

Liu, J., Li, J., Jang, S.S., Zhao, Y. (2022). Understanding tourists' environmentally responsible behavior at coastal tourism destinations. *Mar. Policy, 143*, 105178. [http://dx.doi.org/10.1016/j.marpol.2022.105178]

Mammo, F.K., Amoah, I.D., Gani, K.M., Pillay, L., Ratha, S.K., Bux, F., Kumari, S. (2020). Microplastics in the environment: Interactions with microbes and chemical contaminants. *Sci. Total Environ., 743*, 140518. [http://dx.doi.org/10.1016/j.scitotenv.2020.140518] [PMID: 32653705]

Mathieu-Denoncourt, J., Wallace, S.J., de Solla, S.R., Langlois, V.S. (2015). Plasticizer endocrine disruption: Highlighting developmental and reproductive effects in mammals and non-mammalian aquatic species. *Gen.*

Comp. Endocrinol., 219, 74-88.
[http://dx.doi.org/10.1016/j.ygcen.2014.11.003] [PMID: 25448254]

Mihai, F.C., Gündoğdu, S., Markley, L.A., Olivelli, A., Khan, F.R., Gwinnett, C., Gutberlet, J., Reyna-Bensusan, N., Llanquileo-Melgarejo, P., Meidiana, C., Elagroudy, S., Ishchenko, V., Penney, S., Lenkiewicz, Z., Molinos-Senante, M. (2021). Plastic pollution, waste management issues, and circular economy opportunities in rural communities. *Sustainability (Basel), 14*(1), 20.
[http://dx.doi.org/10.3390/su14010020]

Newman, S., Watkins, E., Farmer, A., Brink, P. T., Schweitzer, J. P. (2015). *The economics of marine litter.*
[http://dx.doi.org/10.1007/978-3-319-16510-3_14]

Pain, A., Hansen, K. (2019). *Rural Development.* London, UK: Routledge.
[http://dx.doi.org/10.4324/9781315773643]

Pásková, M., Štekerová, K., Zanker, M., Lasisi, T.T., Zelenka, J. (2024). Water pollution generated by tourism: Review of system dynamics models. *Heliyon, 10*(1), e23824.
[http://dx.doi.org/10.1016/j.heliyon.2023.e23824] [PMID: 38226237]

PlasticsEurope. (2019). *Plastics—the facts 2019: An analysis of European plastics production, demand and waste data.* PlasticsEurope. https://plasticseurope.org/wp-content/uploads/2021/10/2019-Plastics--he-facts.pdf

Saifi, A., Susanto, H. (2024). Analysis of tourists' behavior in disposing of garbage in the somba fort area opu makassar analisis perilaku wisatawan dalam membuang sampah di kawasan benteng somba opu makassar. *Pendidikan Dan Humaniora, 8*(1).
[http://dx.doi.org/10.36526/js.v3i2.3579]

Saifi, A., Susanto, H. (2024). Analysis of tourists' behavior in disposing of garbage in the Somba Fort area: Opu Makassar (Analisis perilaku wisatawan dalam membuang sampah di kawasan Benteng Somba Opu Makassar). *Pendidikan Dan Humaniora, 8*(1)
[http://dx.doi.org/10.36526/js.v3i2.35799]

Sutter, J.D. (2016). *How to stop the sixth mass extinction.* https://edition.cnn.com

Wiedinmyer, C., Yokelson, R.J., Gullett, B.K. (2014). Global emissions of trace gases, particulate matter, and hazardous air pollutants from open burning of domestic waste. *Environ. Sci. Technol., 48*(16), 9523-9530.
[http://dx.doi.org/10.1021/es502250z] [PMID: 25019173]

Yhdego, M. (1995). Urban solid waste management in Tanzania Issues, concepts and challenges. *Resour. Conserv. Recycling, 14*(1), 1-10.
[http://dx.doi.org/10.1016/0921-3449(94)00017-Y]

Zhao, S., Zhu, L., Li, D. (2015). Characterization of small plastic debris on tourism beaches around the South China Sea. *Reg. Stud. Mar. Sci., 1*, 55-62.
[http://dx.doi.org/10.1016/j.rsma.2015.04.001]

Ziale, J., Quoquab, F. (2023). Plastic pollution in the tourism sector: Documenting its present research status. In *Socially responsible plastic.* Emerald Publishing Limited.
[http://dx.doi.org/10.1108/S2043-052320230000019002]

Ziani, K., Ioniță-Mîndrican, C.B., Mititelu, M., Neacșu, S.M., Negrei, C., Moroșan, E., Drăgănescu, D., Preda, O.T. (2023). Microplastics: A real global threat for environment and food safety: A state of the art review. *Nutrients, 15*(3), 617.
[http://dx.doi.org/10.3390/nu15030617] [PMID: 36771324]

Diener, A., Mudu, P. (2021). How can vegetation protect us from air pollution? A critical review on green spaces' mitigation abilities for air-borne particles from a public health perspective - with implications for urban planning. *Sci. Total Environ., 796*, 148605.
[http://dx.doi.org/10.1016/j.scitotenv.2021.148605] [PMID: 34271387]

Eisfeld-Pierantonio, S.M., Pierantonio, N., Simmonds, M.P. (2022). The impact of marine debris on

cetaceans with consideration of plastics generated by the COVID-19 pandemic. *Environ. Pollut., 300*, 118967.
[http://dx.doi.org/10.1016/j.envpol.2022.118967] [PMID: 35134431]

Evode, N., Qamar, S.A., Bilal, M., Barceló, D., Iqbal, H.M.N. (2021). Plastic waste and its management strategies for environmental sustainability. *Case Studies in Chemical and Environmental Engineering, 4*, 100142.
[http://dx.doi.org/10.1016/j.cscee.2021.100142]

Gálvez-Ontiveros, Y., Páez, S., Monteagudo, C., Rivas, A. (2020). Endocrine disruptors in food: impact on gut microbiota and metabolic diseases. *Nutrients, 12*(4), 1158.
[http://dx.doi.org/10.3390/nu12041158] [PMID: 32326280]

Gibovic, D., Bikfalvi, A. (2021). Incentives for plastic recycling: how to engage citizens in active collection. empirical evidence from Spain. *Recycling, 6*(2), 29.
[http://dx.doi.org/10.3390/recycling6020029]

Greenwood, S.C., Walker, S., Baird, H.M., Parsons, R., Mehl, S., Webb, T.L., Slark, A.T., Ryan, A.J., Rothman, R.H. (2021). Many Happy Returns: Combining insights from the environmental and behavioural sciences to understand what is required to make reusable packaging mainstream. *Sustainable Production and Consumption, 27*, 1688-1702.
[http://dx.doi.org/10.1016/j.spc.2021.03.022]

Higuchi, C., Isobe, A. (2024). Reduction scenarios of plastic waste emission guided by the probability distribution model to avoid additional ocean plastic pollution by 2050s. *Mar. Pollut. Bull., 207*, 116791.
[http://dx.doi.org/10.1016/j.marpolbul.2024.116791] [PMID: 39121802]

Islam, Md. (2024). Impacts and mitigation strategies of microplastic pollution. In N. A. Khan & L. Singh (Eds.), *ACS symposium series* (Vol. 1482, pp. 157–179). American Chemical Society.
[http://dx.doi.org/10.1021/bk-2024-1482.ch007]

Javaid, M., Haleem, A., Singh, R.P., Suman, R., Gonzalez, E.S. (2022). Understanding the adoption of Industry 4.0 technologies in improving environmental sustainability. *Sustainable Operations and Computers, 3*, 203-217.
[http://dx.doi.org/10.1016/j.susoc.2022.01.008]

Kumar, R., Verma, A., Shome, A., Sinha, R., Sinha, S., Jha, P.K., Kumar, R., Kumar, P., Shubham, , Das, S., Sharma, P., Vara Prasad, P.V. (2021). Impacts of plastic pollution on ecosystem services, sustainable development goals, and need to focus on circular economy and policy interventions. *Sustainability (Basel), 13*(17), 9963.
[http://dx.doi.org/10.3390/su13179963]

Lăzăroiu, G., Ionescu, L., Uță, C., Hurloiu, I., Andronie, M., Dijmărescu, I. (2020). Environmentally responsible behavior and sustainability policy adoption in green public procurement. *Sustainability (Basel), 12*(5), 2110.
[http://dx.doi.org/10.3390/su12052110]

Leslie, E. (2023). Synthetic beginnings. In *E. Leslie, The rise and fall of Imperial Chemical Industries* (pp. 1–59). Springer Nature Switzerland.
[http://dx.doi.org/10.1007/978-3-031-37432-6_1]

Macheca, A.D., Mutuma, B., Adalima, J.L., Midheme, E., Lúcas, L.H.M., Ochanda, V.K., Mhlanga, S.D. (2024). Perspectives on plastic waste management: challenges and possible solutions to ensure its sustainable use. *Recycling, 9*(5), 77.
[http://dx.doi.org/10.3390/recycling9050077]

Moschen-Schimek, J., Kasper, T., Huber-Humer, M. (2023). Critical review of the recovery rates of construction and demolition waste in the European Union – An analysis of influencing factors in selected EU countries. *Waste Manag., 167*, 150-164.
[http://dx.doi.org/10.1016/j.wasman.2023.05.020] [PMID: 37267878]

Narancic, T., Cerrone, F., Beagan, N., O'Connor, K.E. (2020). Recent advances in bioplastics: application

and biodegradation. *Polymers (Basel), 12*(4), 920.
[http://dx.doi.org/10.3390/polym12040920] [PMID: 32326661]

Narayan, P., Kar, S., Jha, K.N. (2024). In-depth analysis of risks and their impacts on solar power project development: public and private contractors' perspective. *Journal of Legal Affairs and Dispute Resolution in Engineering and Construction, 16*(4), 05024005.
[http://dx.doi.org/10.1061/JLADAH.LADR-1174]

Ncube, L.K., Ude, A.U., Ogunmuyiwa, E.N., Zulkifli, R., Beas, I.N. (2021). An Overview of Plastic Waste Generation and Management in Food Packaging Industries. *Recycling, 6*(1), 12.
[http://dx.doi.org/10.3390/recycling6010012]

Nkomo, W. (2024). *Development of a low-cost smart monitoring and control system for water management in rural wastewater treatment.*
[http://dx.doi.org/10.25381/CPUT.27628464.V1]

Obersteiner, G., Gollnow, S., Eriksson, M. (2021). Carbon footprint reduction potential of waste management strategies in tourism. *Environ. Dev., 39*, 100617.
[http://dx.doi.org/10.1016/j.envdev.2021.100617] [PMID: 34513580]

Piubello Orsini, L., Leardini, C., Danesi, L., Guerrini, A., Frison, N. (2023). Circular economy in the water and wastewater sector: Tariff impact and financial performance of SMARTechs. *Util. Policy, 83*, 101593.
[http://dx.doi.org/10.1016/j.jup.2023.101593]

Stephens, N., Hurley, R.A., Kimmel, R., Bridges, W., Frierson, A., Darby, D., Skaggs, J., Albro, M. (2023). The environmental impacts of caesar salad packaging. *Sustainability (Basel), 15*(13), 10260.
[http://dx.doi.org/10.3390/su151310260]

Vashishth, T.K., Sharma, V., Agarwal, S., Gupta, N., Singh, S., Sharma, R. Embracing the current challenges in sustainable waste management in Tourism and Hospitality. *Advances in Hospitality, Tourism, and the Services Industry.* IGI Global.
[http://dx.doi.org/10.4018/979-8-3693-7898-4.ch011]

CHAPTER 12

The Impact of Artificial Intelligence in Environmental Pollution–A Dwelling Aspect for a Better Sustainable Tomorrow

Geetha Manoharan[1,*] and **Sanjeev Kumar**[2]

[1] *School of Business, SR University, Warangal, Telangana, India*

[2] *School of Hotel Management & Tourism, Lovely Professional University Phagwara, Punjab, India*

Abstract: Because the transformative power of Artificial Intelligence (AI) continues to fuel innovation, concerns are growing about its negative impact on the environment. In this article, we examine the sustainability issues surrounding AI, focusing on related problems such as carbon emissions, digital waste, and environmental impacts. Through proactive actions and ethical practices, we must learn how civilization can effectively cope with these concerns and strive to build a sustainable world where AI and nature conservation coexist harmoniously. From an environmental perspective, AI presents both positive and negative possibilities. On the positive side, AI can aid in alleviating most of the major climate problems; on the other hand, it can also exacerbate the issue. By transferring in advance, we must continue to utilize this powerful gadget in the proper ways: reducing ocean pollution, aiding conservation efforts, and enhancing recycling techniques. However, we also want to find ways to power the machines with less environmental impact, and that can be accomplished without undue delay. AI is hailed as a game-changer, but under its transformative capability lies a pressing concern: its environmental impact. The improvement, protection, and use of AI generation technologies have a significant carbon footprint. In this chapter, we delved into the hidden environmental predicament surrounding AI and uncovered the urgent need for proactive measures. A sustainable future can be secured, and the dangers AI poses to our world community can be mitigated by being aware of the risks and acting responsibly.

Keywords: Artificial Intelligence, Environment, Safety Management, Sustainability, Pollution.

* **Corresponding author Geetha Manoharan:** School of Business, SR University, Warangal, Telangana, India; E-mail: geethamanoharan1988@gmail.com

Sanjeev Kumar, Mohammad Badruddoza Talukder & A.K. Haghi (Eds.)

INTRODUCTION TO ENVIRONMENT

Artificial intelligence extends far beyond high-profile chatbots; machine learning now powers a diverse array of ubiquitous technologies, from GPS navigation systems to the algorithms that curate social media feeds. But as AI's reputation continues to grow, more researchers and specialists are noting the environmental cost. Education and AI tool usage require a significant amount of computational power, and the consequent carbon dioxide emissions are just one of the ways AI impacts the climate. Its impact on the environment is much greater than its carbon footprint. Artificial intelligence (AI) can have a profound effect on the environment, both positive and negative. The advancement and use of AI have changed many aspects of our lives, as well as our relationship with the world. With its ability to analyze unprecedented volumes of facts, examine patterns, and make decisions in real-time, AI can be utilized to optimize energy efficiency, reduce waste, and enhance the aesthetic appeal of sustainable practices. But the negative environmental effect of AI is also a contributing factor to this scenario. Bhamidipati, R.T. (2025).

The significant environmental impact of AI is evident in several areas. A considerable advantage of AI is that it can optimize power consumption and reduce waste. For instance, machine-learning algorithms can review data from smart grids to optimize power consumption in real-time, reducing the need for fossil fuel-based overall strength generation. Such initiatives may reduce greenhouse gas emissions and curb the impact of climate change. Akbar, N.S. (2025).

AI can also expand and enforce environmentally sustainable practices in sectors like agriculture, forestry, and transportation. For instance, precision agriculture can help farmers reduce their use of fertilizers and pesticides, resulting in healthier crops and lower environmental pollution. Additionally, AI-driven forestry management can help ensure that forests are sustainably harvested with minimal impact on the surrounding environment. In transportation, AI can help optimize routes and reduce gasoline consumption, which results in fewer emissions and cleaner air. Naga, H., Bhaskar, V., & Mayil, V.V. (2025).

Environmental Pollution

M, A. In the 2025 step, researchers highlight that air pollutants continue to be a significant environmental and public health concern, and that particles are the primary contributors to air quality degradation. This study aims to investigate various methods for assessing perception and predicting air quality using statistical procedures, time series analysis, and system modeling strategies. Traditional fashion, including regression analysis, ARIMA, and time series decomposition, was typically applied to the first class of air ratings. However, advances in Artificial Intelligence (AI) have introduced specialized predictive models and Support Vector Machines (SVM), Deep Convolutional Neural Networks (DCNN), and Long Short-Term Memory (LSTM) networks. Furthermore, we examine the effects of meteorological elements on the distribution of pollutants and the effectiveness of strategies for urban greening to reduce air pollutants. The results demonstrate that hybrid AI models, which combine deep learning and statistical methods, possess excellent predictive capabilities and offer promising techniques for real-world applications and informed decision-making. AI can be used to occupy new substances in unique homes. It also enhances the degree of reduction in strength or weight available for utilization during production, up to an aerospace level. The materials can be based on renewable assets that reduce reliance on fossil fuels and mitigate the environmental implications of production. These reforms will enable you to identify and address ecological problems more effectively. For instance, you use AI to visualize and project anticipations. This technology allows us to prepare for storm activities and minimize our social and environmental impact. Through AI, we analyze and publish ecological data, including air and water quality information, to identify areas for improvement and aim for solutions. A., Madhuri, J.V., Saravanan, D., Vijayakumar, T., Manimegalai, D.V. The development of AI algorithms takes a lot in terms of resources. The biggest fear is the power of electricity to train and run AI algorithms. Training AI models used in education can consume a massive amount of computer energy. This processing takes immense electricity. This intensity is often generated by fossil fuels, resulting in a surge of greenhouse gas emissions.

Another issue is the ability to compress AI to address environmental challenges. For instance, AI-boosted automation will result in high-speed recording and reduced waste in industries such as e-commerce, where quick and widespread delivery is made the norm. Similarly, AI-based agriculture is a green health initiative. Qin, Z., Wei, B., Gao, C., Chen, X., Zhang, H, Wong, C.U. (2025). The easiest way to run AI algorithms is to use the statistics on which they have been trained, and this data distortion can lead to biased decisions. For example, suppose the AI-generated rules are trained with statistics that prioritize monetary gains over environmental security. In that case, decisions can be made that prioritize long-term financial benefits over short-term ecological concerns. With the help of renewable energy assets from solar or wind power, this can contain and make the AI model work, Manoharan, G., Ashtikar, S. P., & Nivedha, M. (2024). Applying AI in a manner that encourages sustainable behavior and minimizes environmental footprint is another alternative. We can utilize AI as a model to maximize recycling technology and reduce food waste in delivery chains. The increasing need for branding with new and sophisticated AI technology has led to the generation of high volumes of digital waste. E-waste disposal is a serious environmental issue, as it leads to soil contamination, water pollution, and air pollution. Miles should develop responsible and sustainable methods for disposing of and recycling digital waste. This consumer boom could contribute to vegetable assets fatigue, which has a significant environmental impact on flips. As a result, it is crucial to ensure that AI is utilized to enhance resource efficiency and maintenance.

AI in Environmental Pollution

The benefits of Artificial Intelligence (AI) in environmental monitoring include accurate disaster forecasting, pollutant identification, and comprehensive monitoring of air and water quality. David B. Olawade, Ojima Z. Wada (2024). The piece presents various environmental monitoring perspectives, highlighting the shortcomings of conventional approaches and their associated costs. Numerous large-scale AI applications in environmental monitoring have been identified to contribute to efficient ecological management. A management. AI technology enhances environmental monitoring by facilitating a better understanding, prediction, and mitigation of environmental hazards. Yet, the overall potential of AI reveals challenges in the ecological field, including a lack of specialist AI professionals and statistical difficulties. These problems are more significant in regions where technological infrastructure is developing. In this research, paintings serve as an active documentation of governance activities, aided by the government, with a record to be maintained. Owing to such issues, however, the future of AI remains hopeful in environmental monitoring. This is because developments in AI algorithms, data recording techniques, and processing capacity are likely to further increase the accuracy and efficiency of pollution monitoring and management. Remain true to the appeal of current climates. The potential impact of an AI environment is just one aspect of the potential for artificial intelligence to be developed through global initiatives aimed at mitigating climate change. Climate change is threatening one of the planet's coldest climate regions, but fortunately, the potential impact of AI on the environment may help mitigate this scenario. This is because AI can track the melting of icebergs 10,000 times faster than human observation (Poongodi & Manoharan, 2025). Especially during the latest disaster, the use of AI and social media in technology-based recovery efforts has been carried out in a commendable manner.

Map-Based: While attempts to enhance the cultivation of new and trendy plants, like soybeans, and even city-building at international events, like the Olympics, tend to result in design, the good news is that AI in space assists humans in enhancing the largest areas of the country. In 2021, 85% of plastic waste was exported to the US. However, facilities can address this issue with new software that enables them to monitor and sort incoming materials more effectively, resulting in better recycling. The trash may then be eliminated. This case study illustrates how AI can prevent future plastic pollution.

Preparation against natural disasters: Hurricanes, tornadoes, tsunamis, and earthquakes - no matter what form the natural disaster takes, these devastating events created by Mother Nature always have a lesson or two to impart to their descendants. AI minimizes the long-term effects of natural disasters by providing advance notice when they are likely to occur. This leaves the community prepared well in advance.

Supply Chain Management: AI enables the decarbonization of the supply chain and enhances overall business efficiency by utilizing various algorithms, preventing disruptions, and monitoring all operational activities. AI can accomplish such activities more precisely and avoid potential human errors. A couple of years ago, I wrote a paper that discussed the advantages of AI technologies in the chemical industry. AI enhances energy efficiency, reduces energy consumption, and supports environmental monitoring and management. B. It helps to avoid air pollution. AI also facilitates process and supply chain optimization for reduced ecological effects. Grahlov UE, J. (2025).

Types of Help that AI can Give for Environmental Pollution

Tracking and Preventing Deforestation

Still, experts are now using AI to challenge this difficulty, and here's how:

Deforestation detection: Scientists can now combine AI with satellite images to identify which areas are affected by deforestation. For example, the Scottish-based company Area Intelligence has mapped more than 1 million hectares of land in 30 private countries using this method. Aside from raising awareness of deforestation, this gadget can also help prevent illegal logging activities worldwide.

Prevention of deforestation: We will rely on ongoing monitoring to prevent deforestation. We must also create awareness to prevent it from continuing, and that is why researchers are employing AI to develop predictive models that can identify areas at risk, primarily based on variables such as land-use patterns, infrastructure development, and population density. These figures can then guide human beings within the enterprise in spearheading conservation activities and addressing areas that require assistance.

Making Plastic Recovery More Efficient

The Ellen MacArthur Foundation estimates that by 2050, there will be more plastic in our oceans than fish. In addition, to make matters worse, about 85% of plastic waste ends up in landfills. Removing this trash is an enormous undertaking. That's why professionals within the company are applying AI to green the technique.

Some of the Flags include:

Photo compliance, which detects bare hands and feet in photos, enables us to ensure that the waste series team meets health and safety compliance standards by providing shoes and gloves. The number of workers wearing shoes rather than sandals has increased dramatically since this option became available. “Image weight mismatch” warns us if the weight displayed on the size and the weight manually entered don't coincide, which is essential in ensuring our collection data is accurate. The system tracks weight outliers, reporting any weights outside the normal range for the collection hub. Identifying this challenge is crucial for ensuring transparency and accuracy in our data collection, as noted by Manoharan and Kumar (2025).

Removing Plastic from the Ocean

It's estimated that there are 21,000 pieces of plastic floating in the ocean. Fortunately, agencies are trying to eliminate this waste from the oceans; however, finding and tracking plastic contamination in the sea is simpler said than done. This is where AI steps in.

Certain organizations are going greener with cleanups by using satellite imagery and device readings to learn about and become aware of ocean plastic particles. For instance, the Sea Cleanup is a non-profit organization that utilizes AI to identify and monitor plastic contaminants within the gigantic Pacific Garbage Patch (a massive collection of plastic debris floating in the Pacific Ocean). The company utilizes a fleet of autonomous vessels to collect information on the position and length of plastic rubbish, assisting in the production of a detailed map of ocean trash in certain remote regions. However, AI is capable of doing more than just charting plastic pollution—it could also help expand the new generation of cleanups. For instance, scientists at the University of California, Berkeley, have developed a robot that can swim in the ocean and collect plastic particles. The robot comes equipped with a digital camera and a suction device, enabling it to identify and remove plastic waste from the water.

Decarbonizing Industries

The emissions around the globe have continued to rise over the past century, in tandem with the growth of businesses. In reality, we now release more than 35 billion tons of carbon dioxide (CO_2) per year, compared to 6 billion in 1950. Since commercial industries produce approximately 30% of global greenhouse gases, scientists are considering a method of rapidly decarbonizing them—and here are some ways AI is assisting them with this endeavor: Greening supply chains: AI can study intricate supply chains and guide them to greener by means of pointing out important possibilities that involve optimizing transportation routes, strengthening strength performance, reducing packaging squander, and reducing stock stages. Gomez, ok.S., Nair, A.S., AbhinandC, B., JishnuK, S., RahulK, S., & Binny, S. (2024). Improving building performance: Companies can leverage AI to monitor and regulate building systems, streamlining power consumption, heating, cooling, and lighting devices. Devices that learn about algorithms can also verify occupancy patterns and adjust strength usage accordingly.

Analyzing transportation plans: AI can be applied to path-making plans, visitor management, and fleet management to minimize carbon footprints. The software will take into account considerations such as the status of site visitors, fuel usage, and other modes of transportation. Kaur, A. (2025). Optimizing electricity systems: AI algorithms can analyze vast amounts of data from power structures, optimizing technology, distribution, and intake. This could help global regions transition away from fossil fuels to cleaner, renewable energy sources, ultimately reducing a significant amount of carbon emissions.

Prevent Climate Disasters

Data indicate that the extent of climate-related issues has tripled over the last 30 years, which aligns with the boom in global emissions. Climate disasters, including monsoons, floods, hurricanes, and droughts, are resulting in extensive harm to communities worldwide. Experts are now turning to AI to gain a better understanding of this changing weather and its impact on vulnerable groups. For example, various UN organizations are assisting communities in Burundi, Chad, and Sudan—regions prone to weather exchange breakdowns. This challenge, driven by AI, will examine historical environmental changes near displacement hotspots and provide future projections that may help these groups act and adjust to the transforming climate. AI is also involved in the UN Secretary-General's Early Warnings for All initiative—a motion plan that seeks to ensure everyone is protected from perilous climate, water, or weather events through early warning systems up until 2027. Another instance of applying AI to forecast weather interruptions includes the Brazilian firm Sipremo. The company, which is now expanding into businesses such as insurance, energy, logistics, and sports, intends to utilize its software to help agencies and governments prepare more effectively for climate change.

Tracking Air Pollution

According to the World Health Organization, the World Bank estimates that the greatest environmental-related cause is for an estimated 8.43 million premature deaths within 12 months. For years, scientists have been tracking air pollution phases through tracking stations, which collect a vast amount of data. In simple words, it is painstakingly time-consuming. And that's exactly the reason why Cornell College engineers have developed a streamlined version to monitor air pollutants. The gadget feeds data such as traffic conditions, road maps, and climate, applying artificial intelligence rules to examine various circumstances involving traffic-borne air pollution. This technology enables city planners and government health officials to gain a better understanding of the air people breathe in a city.

Recycling Plastic Waste

According to the OECD, researchers estimate that only 9% of the industry's plastic can be recycled. This is an international issue that needs to be addressed if we are to steer towards a circular economy. These are the hip AI advancements that are making the recycling business turn over:

Smart recycling boxes, which feature AI technology, have the potential to encourage individuals to recycle the right devices by utilizing sensors and cameras that detect the type of waste being discarded. Waste Net is a great example of the success of this technology, achieving a prediction rate of 97% and sorting waste into six categories: paper, cardboard, glass, metal, plastic, and other materials. Supriya Rai, D., Nadeem, T., Bhansali, P., Sanghvi, B., Bafna, H., & Lunkkadd, M. (2025).

Green waste recovery: London-based Company Greyparrot created an AI system that examines incoming waste at recycling plants to recover and recycle additional waste fabric. The business alone monitored 32 billion waste devices within 67 waste streams in 2022. Recycling robots: London-headquartered, entirely AI-driven company Recycleye has developed a robot selection tool to simplify dry and mixed recyclable sorting. The company now has 30 AI-driven robots hired in the UK and Ireland, with 19 potentially already connected, as well as additional ones mounted across other areas of Europe.

Reducing Fast Fashion

As rapid fashion trends continue to benefit the industry's reputation, the amount of waste it generates piles up excessively. The clothing industry currently generates 92 million tons of textile waste annually. There is still a long way to go before we achieve sustainable business. Still, some organizations have begun making changes in the right direction — and AI is assisting them on this journey. AI can enhance operations and supply chain practices by streamlining inventory management and forecasting revenue, primarily based on historical statistics. This aspect can be instrumental in preventing manufacturers from overstocking clothing and sending wasted materials to landfills.

Frenetic fashion is detrimental to the world in numerous ways, but back-to-store gadget returns are among the worst of them. As much as 24 million metric tons of CO_2 emissions are linked with e-commerce returns annually. For example, AI 'try-on' technology allows clients to view how clothes will look on them without actually trying them on, saving return expenses and emissions. Several groups, such as prescription eyewear retailers Warby Parker and Amazon, have already taken this route.

Supporting the Transition of Renewable Energy

Since 1950, the use of fossil fuels has increased by approximately eight times, contributing to global warming and climate change. That is why enterprise specialists are urging nations to transition to cleaner energy sources, such as solar, wind, or hydroelectric power. In fact, artificial intelligence is helping us make this transition. This is not a new collaboration for both. The most recent projection suggests that AI is already in more than 50 unique applications in the strength gadget. One of the most frequent applications of AI in the electricity sector is to enhance supply and demand forecasts. Renewables such as wind and solar are intermittent (since it's not always windy or sunny), which has in the past constrained their growth. However, now AI can help researchers gain a superior understanding of what renewable energy sources are most likely to vary. For example, researchers can predict wind electricity output using climate patterns and information collected from nearby wind farms. AI algorithms can also determine the optimal time to retain electricity—another major project for the enterprise. Therefore, we have been utilizing all the above-mentioned blessings throughout the years, and a significant shift has occurred, discussing the arrival of artificial intelligence in environmental elements as well.

AI Measures towards Environmental Pollution

Computer systems of AI need both strength and energy to function. Energy technology is responsible for discharging pollutants, especially in areas where fossil fuels control power generation. The intake of strength is affected by the combustion of fossil fuels, which produce greenhouse gas emissions and contribute to climate change. The combustion of fossil fuels also releases pollution that contributes to air and water pollution, respiratory illnesses, and acid rain. Fossil fuel extraction and power infrastructures can encroach on ecosystems and contribute to environmental degradation. The shift to using renewable energy sources, such as the sun and wind, and the establishment of energy-efficient behaviors can avert these adverse outcomes. The manufacturing, transportation, maintenance, and disposal of these hardware components, *i.e.*, servers and data centers, require additional electricity use and considerable amounts of natural resources. B. Cobalt, silicon, gold, and numerous other metals, Alham mites, M. O, Kaplan, F., & Rufai Ogle, S.B. (2025). The production and mining of such metals used in AI hardware result in soil erosion and pollution. Most electronic devices are not well recycled, primarily because of e-waste, which can also lead to health issues. The chemicals applied in such devices, if not disposed of properly, may pollute the soil and water.

Challenges in AI towards Environmental Sustainability

Right here are several of the methods AI may demonstrate to be harmful to the planet:

- **Disposal and Electronic Waste**

AI should have more generations and computer systems to be taking strolls than typical, which is, on the verge of its existence cycle—AI ought to generate more electronic waste than previous computer systems deployed in commercial centers. Additional strength intake: AI frameworks can assist us in perceiving things we cannot see with our own eyes; yet, the cost of this increased information can burden the Earth, as additional power is needed to operate AI frameworks (many of which may need to consume considerable energy to demonstrate their capabilities constantly). Hence, AI must emit more carbon, primarily if the AI architectures utilized across organizations aren't running on renewable electricity resources.

- **Should create a lack of incentive**

Although AI can indeed help with the development of an enterprise's climate plan, it cannot implement that plan for the given organization. Thus, AI might hurt the environment by generating too many ideas, to the extent that it prevents the development of a coherent action plan. AI could put pressure on our power grids, as it has been established that new technological developments, such as AI, often lead to power shortages. That being said, the idea is that the more we move away from using fossil fuels to power our homes, the more energy will be required to reinforce them, and that will run short if the entire amount of power is going towards powering AI or conducting research on AI.

- **Essential facts about storage**

AI cannot operate without adequate data storage, which produces its own carbon footprint, including the maintenance and the energy used to cool the storage of warm records. Capability AI Miscalculations—AI is an overreliance on mundane tasks that nobody wants to accomplish when a computer can achieve them, such as spell-checking or requesting the weather outside. But, since AI does not possess the personal judgment that humans do, its algorithms may misestimate the effect a natural disaster may have on a community. The affected area may not receive the necessary resources. To mitigate the potential negative environmental impact of AI, it is essential to establish robust and enforceable policies and standards for the development, deployment, and disposal of AI technology. Those formulating policies should consider the potential environmental impacts of AI and promote green practices. For example, regulations must mandate that agencies employ energy-efficient AI models or recycle virtual waste.

In addition, it's much more imperative to educate the majority of the population about the capability to comprehend the ecological impact of AI and the market. Individuals can contribute significantly towards the development and application of AI generation through informed decision-making and promoting sustainable means.

Prevention and Safety Management of Environmental Pollution

Here are some methods we ought to apply to mitigate the impact of a catastrophic AI environmental influence:

- **Additional green electricity programs:**

AI frameworks should reduce their energy consumption if their programs align with the energy efficiency needs of today. Freshly refined networks and the development of additional green electricity programs may make this possible.

Middle improvement facts: choosing more energy-efficient record buildings or seeking to enhance the current statistics within your firm's AI device will aid in minimizing your firm's AI carbon footprint. Some ideas for selecting a higher statistic for the middle storage device include utilizing centers that lease renewable energy, streamlining data center infrastructure and cloud computing systems to enhance energy efficiency, and minimizing the carbon footprint of AI schooling and deployment. This includes employing methods such as dynamic resource allocation, server virtualization, and the integration of renewable energy sources.

- **Higher-level AI education and statistics**

AI works a lot like the saying “monkey see, monkey do.” This means that adequate AI training must be completed so that AI frameworks are free from bias when providing individuals with varied statistics. Thus, focusing on specific fact examples and refraining from repetitive series of information or input is also important because it helps prevent and mitigate AI bias. Standard AI shouldn't have a significant environmental impact. Still, we can significantly reduce the likelihood of such an impact by working to make AI more energy-efficient and aware of when it is best to be utilized. If we proceed with this idea, the benefits to be gained from AI's environmental impact could help us manage the global endeavor of climate change.

DISCUSSION

Pushing AI forward so that it draws power from a cleaner source of energy goes hand in hand with the shift of the entire economic system towards renewable energy sources—that is, sources other than oil, gas, or coal. The research into how one can transition towards making this change appears to continue; yet, seeking even more efficient sources of power for manufacturing, production, and transportation is evidently a key area to explore. We conjecture that the primary coverage challenge with AI currently is learning how to regulate the fast pace of innovation, particularly in the individual sector, as well as continued safeguarding of the general population from other risks associated with artificial intelligence, including privacy, discrimination, and control. Nugroho, N.M., Suharyono, ok.P., Hadiningrat, S., (2025).

GPT holds out the promise of a huge productivity boost. It's a technology designed to revolutionize how we live and work. Therefore, I believe AI requires some time to evolve further before we can expect carbon neutrality in AI computing. AI is still in its early stages of development, and we hope it will become greener than it is today through ongoing improvements in both software and hardware.

We are entering an era where we must do more with less. While AI can be used responsibly, it can also be utilized to help resolve many of society's issues and problems. Its applications are being used in an increasingly larger number of private sector and government initiatives that claim to have or not have a demand for public decency. What presents us with hope and optimism about AI, therefore, is its likely higher potential to help tackle difficult coverage matters that are becoming increasingly challenging. Era had the sole response; yet, improving the fact base for choice making is important.

Yet, to utilize AI properly and to modify it well also means we need to take care of the matter that regulatory brilliance is frequently "people's brilliance." In other words, we need human beings to utilize these AI tools wisely and effectively, recognize how they are not capable of performing work appropriately, and market them publicly appropriately. The utility of AI requires a contemporary spark from actual human beings to identify areas where this generation can be utilized most effectively, and then to plan it out with caution and vigilance. The individuals who have sounded the alarm on AI and its risks deserve commendation, and their concerns should be taken seriously and listened to. We wish to continue with management systems and auditing processes that are meant to uncover problems that drive costs upward—and then correct them.

CONCLUSION

The ecological consequences of the meteoric rise of AI are a cause for real concern and require urgent attention. While the allure of AI's promise is easy, we need to face its dire impact squarely. By shedding light on the underlying environmental dilemmas surrounding AI, humankind can facilitate an informed discussion and drive the adoption of responsible practices.

In fact, integrating AI in environmental observation offers paradigm-shifting advantages across various sectors, including top-notch soil, water, and air monitoring, as well as traffic regulation and carbon footprint monitoring. Chen, Z. (2025). These developments significantly contribute to environmental security, individual health, and sustainable development. AI's ability to provide accurate predictions and live monitoring enhances the efficacy and effectiveness of environmental management techniques. But it's much more crucial to forego the capability shortcomings associated with the rollout of AI technologies. One significant issue is the substantial power burden imposed by record centers and the delivery chain, which contributes to enhanced greenhouse gas emissions.

The scarcity of skilled professionals in the environmental sector, particularly in the Global South, presents another challenge to fully leveraging AI's potential. Records are given proper entry into control, and privacy issues must be addressed to prevent the abuse of AI structures for greater advantage, including marketplace manipulation or the exploitation of disaster prediction. Robust information governance regimes are crucial in mitigating these risks and ensuring the equitable and fair utilization of AI technology. Despite the concerning circumstances, the future of AI in environmental monitoring holds great promise. Future research should place a high emphasis on examining the ethical considerations and measuring the ecological impact of AI technology. Constructing standardized methodologies to assess the environmental impact of AI systems and examining the moral implications of their use will facilitate the responsible and sustainable application of AI.

In summary, the impact of AI on the environment is substantial, encompassing both positive and negative effects. AI can sell electricity performance, reduce waste, and increase sustainable practices. But it also has the potential to contribute to environmental degradation and natural property depletion. It's much more critical to act to prevent the horrible environmental impacts of AI and encourage responsible and sustainable utilization. By doing this, we can ensure that AI leads to a more sustainable and environmentally improved future.

The transformative force in the global mission to reduce environmental pollution will establish sustainable standards for a better planet. The ability to understand and interpret data through visualization has different pathways in predicting disasters across various systems. Through the use of sensors, the quality of air and water can be consistently measured. Since the integration of green environments and AI, there are possible positive aspects that are happening to make the Earth a better place to live in. There are many government and non-government agencies that implement the best feasible practices for creating an environment with good energy levels. Finally, it is human wisdom that should ensure we leave this planet a better place to sustain, with good energy levels and harmony.

REFERENCES

Bhamidipati, R. T. (2025). Innovative strategies for sustainable environmental management: AI and IoT-based approaches. *International Journal of Research and Review in Applied Science, Humanities, and Technology*.
[http://dx.doi.org/10.71143/we2h0g36]

M, A. (2025). A survey on air pollution using deep learning. *International Journal for Research in Applied*

Science and Engineering Technology. *13*(IV).
[http://dx.doi.org/10.22214/ijraset.2025.68610]

Chiu, W.A., Newman, G., Sansom, G., Ye, X., Rusyn, A., Wu, H., Winckelman, T., Rusyn, I. (2025). MyEcoReporter: a prototype for artificial intelligence-facilitated pollution reporting. *J. Expo. Sci. Environ. Epidemiol., 35*(3), 325-329.
[http://dx.doi.org/10.1038/s41370-025-00747-5]

Naga, H., Bhaskar, V., Mayil, V.V. Forecasting traffic air pollution increases from urban transportation using machine learning in metropolitan cities—A review in Proceedings of the 4th International Conference on Sentiment Analysis and Deep Learning (ICSADL) (pp. 1280–1285).
[http://dx.doi.org/10.1109/ICSADL65848.2025.10933458]

Olawade, D.B., Wada, O.Z., Ige, A.O., Egbewole, B.I., Olojo, A., Oladapo, B.I. (2024). Artificial intelligence in environmental monitoring: Advancements, challenges, and future directions. *Hygiene and Environmental Health Advances, 12*, 100114.
[http://dx.doi.org/10.1016/j.heha.2024.100114]

Akbar, N.S. (2025). Leveraging AI for environmental solutions addressing climate change, pollution monitoring and sustainable resource management. *Materials Research Proceedings*.
[http://dx.doi.org/10.21741/9781644903414-97]

Nugroho, N.M., Suharyono, K.P., Hadiningrat, S., Pramudya, F., Wardani, R., Pacuan, J., Raya, K., Putih, K., Gadung, K.P., Jakarta, K. (2025). Analysis of policies and regulations on fire prevention and environmental pollution in the maritime sector: A case study of the Azalia Tanker. *Jurnal Cakrawala Akademika*.
[http://dx.doi.org/10.70182/jca.v1i6.332]

AlHamdani, M.O., Kaplan, F., Rufaioglu, S.B. (2025). Methods used in the detection of heavy metal pollution in soils. *International Journal of Scientific Research and Management, 13*(3), 592-608.
[http://dx.doi.org/10.18535/ijsrm/v13i03.ah01]

Prakash, N., Zielinski, O. (2025). AI-enhanced real-time monitoring of marine pollution: Part 2 — A spectral analysis approach. *J. Mar. Sci. Eng., 13*(4), 636.
[http://dx.doi.org/10.3390/jmse13040636]

Kaur, A. (2025). Artificial intelligence (AI) in traffic management. *International Journal On Science And Technology, 16*(1), 1-6.

SupriyaRai, D., Nadeem, T., Bhansali, P., Sanghvi, B., Bafna, H., & Lunkkadd, M. (2025). Biodegradable sensors in environmental monitoring. *International Journal of Scientific Research in Engineering and Management*.
[http://dx.doi.org/10.55041/IJSREM43525]

Qin, Z., Wei, B., Gao, C., Chen, X., Zhang, H., In Wong, C.U. (2025). SFDformer: a frequency-based sparse decomposition transformer for air pollution time series prediction. *Front. Environ. Sci., 13*, 1549209.
[http://dx.doi.org/10.3389/fenvs.2025.1549209]

Ponnuru, A., Madhuri, J.V., Saravanan, S., Vijayakumar, T., Manimegalai, V., Das, A., Mahalingam, D. (2025). Data-driven approaches to water quality monitoring: Leveraging AI, machine learning, and management strategies for environmental protection. *J. Neonatal Surg., 14*(5S), 664-675.
[http://dx.doi.org/10.52783/jns.v14.2107]

Manoharan, G., Kumar, S. (2025). Embracing the current challenges in sustainable waste management in tourism and hospitality. In: Talukder, M., Kumar, S., Haghi, A., (Eds.), *Application of waste-to-energy innovations in tourism.* IGI Global Scientific Publishing.
[http://dx.doi.org/10.4018/979-8-3693-7605-8.ch002]

Gálová, J. (2025). Using AI to reduce the economic impact of air pollution. In *Conference proceedings* (pp. 124–132). Nitra, Slovak Republic.
[http://dx.doi.org/10.15414/2024.9788055228167.124-132]

Bhuvaneswary, N., Fedrick, A., Alltrin, K., Jaspher, C.J., Sounder, K. (2025). AI-based trash collector boat

for autonomous waterway pollution management. *2025 International Conference on Sustainable Energy Technologies and Computational Intelligence (SETCOM),* 1-5.
[http://dx.doi.org/10.1109/SETCOM64758.2025.10932499]

Chen, Z. (2025). Innovation path of pollution reduction in manufacturing enterprises driven by digital transformation. *Creative Economy, 9*(1), 85-89.
[http://dx.doi.org/10.47297/wspceWSP2516-251917.20250901]

Gomez, K.S., Nair, A.S., Abhinand, C.B., Jishnu, K.S., Rahul, K.S., Binny, S. (2024). AI-enhanced camera systems for real-time identification of expired vehicle pollution and insurance via license plate recognition. *IRJAEM, 2*(12).
[http://dx.doi.org/10.47392/IRJAEM.2024.0528]

Poongodi, K., Manoharan, G. (2025). Michaung Cyclone 2023: Enhancing preparedness and response strategies for future natural disasters. In: Ujager, F., Ali, H., Ahmad, S., (Eds.), *Exploring pillars of sustainability for modern age improvements.* IGI Global Scientific Publishing.
[http://dx.doi.org/10.4018/979-8-3693-5748-4.ch015]

Manoharan, G., Ashtikar, S.P., Nivedha, M. (2024). Unveiling the critical role of artificial intelligence in energy management. In: Riswandi, B., Singh, B., Kaunert, C., Vig, K., (Eds.), *AI applications for clean energy and sustainability.* IGI Global Scientific Publishing.
[http://dx.doi.org/10.4018/979-8-3693-6567-0.ch012]

SUBJECT INDEX

A

B

C

D

E

F

G

H

Sanjeev Kumar, Mohammad Badruddoza Talukder & A. K. Haghi (Eds.)

I

M

N

P

Q

R

S

T

U

www.ingramcontent.com/pod-product-compliance
Lightning Source LLC
LaVergne TN
LVHW070950210926
848482LV00002B/163

* 9 7 9 8 8 9 8 8 1 4 9 1 5 *